The Place-Names
of the Old County of Northumberland

1:
The Cheviot Hills & Dales

Jonathan West

First published by Northern Heritage 2017
Text © Jonathan West

Distributed by
Northern Heritage
Units 7&8 New Kennels, Blagdon Estate, Seaton Burn,
Newcastle upon Tyne NE13 6DB
Telephone: 01670 789 940
www.northern-heritage.co.uk

See our full online catalogue at www.northern-heritage.co.uk

Printed and bound in Great Britain by Martins the Printers.
Berwick-upon-Tweed

British Library Cataloguing in Publishing Data
A catalogue record for this book is available from the British Library.

ISBN 978-0-9957485-1-4

The author and publisher have made every effort to ensure that the information in this book is correct at going to press

All rights reserved.
No part of this book may be reproduced, stored or introduced into a retrieval system, or transmitted
in any form or by any means
(electronic, mechanical, photocopying, recording or otherwise)
without the prior permission of the publisher.

This book was produced with the help of funding from the
Northumberland National Park.

Supported by
Northumberland National Park

Contents

	Acknowledgments and Introduction	iv
1	From Carter Bar to the Ottercops	9
2	From Elishaw to Stagshaw	20
3	From Hexham to Kielder	33
4	From Redesdale to Upper Coquetdale	59
5	From Hepple to the Devil's Causeway	73
6	An excursion around Kidland	79
7	The Cheviot from the East	82
8	The Cheviot from the North	88
9	The Breamish Valley	93
	Further Reading	101
	Technical Terms	102

Flap 1: Place-name elements and abbreviations

Flap 2: Pronunciation

Acknowledgements

Thanks are due first and foremost to Máire West for patient reading of the draft text, and plenty of good ideas (e.g. the possible connection of Welsh *llwm* (Old Irish *lomm*) to LUMSDON LAW). Bill Jones' careful, sympathetic reading of the draft and his encouraging comments were invaluable. Ian Roberts also provided me with a host of helpful suggestions regarding the local historical context, and read a draft. Lord Redesdale generously allowed me to reproduce some details of his ancestor Frances Mitford's wonderful historical views of Redesdale. In addition, I am grateful to Chris Hartnell of Northern Heritage for his unwavering patience over the comparatively few, but still too many, years it has taken to finish the book. Of course, any mistakes and infelicities which remain are mine.

Finally, I should like to record my thanks to Professor Emrys Evans, then Reader in Celtic Studies at the University of Manchester, who first sparked my interest in a serious study of place-names in the autumn of 1971 by pointing out the Brittonic origins of the names of many settlements and natural features in the north-west of England.

There are also a number of books and websites which have turned out to be indispensable. No serious study of Northumberland place-names can ignore Mawer's *The Place-Names of Northumberland and Durham* (Cambridge 1920) and I have made extensive use of this seminal work. Eilert Ekwall's *The Concise Oxford Dictionary of English Place-Names* was also never far from my desk. The English-Scottish border postdates the formation of many place-names, so we can learn a great deal from looking at the place-names of southern Scotland, for which I used two main sources, one from the 1920s (W. J. Watson, *The Celtic Place-Names of Scotland*) and a more recent volume (W. F. H. Nicolaisen, *Scottish Place-Names*). The new 'must-have' work of reference is Watts' *Cambridge Dictionary of English Place-Names*, which contains a judicious and thorough sifting of the evidence.

In trying to make sense of the English data, I have made almost daily use of the online version of the *Oxford English Dictionary* (thanks to Northumberland County Libraries). Regrettably, this service was withdrawn in March 2016 without any form of consultation with or notice to the users.

I would have had a much less detailed knowledge of the history of various places had I not had access to the *Keys to the Past* website, "an exciting Heritage Lottery Funded project launched in 2003, which unlocks the archaeological secrets of County Durham and Northumberland" (http://www.keystothepast.info). There is also a wonderful national project to photograph all the OS squares and the places in them: the Geograph® (http://www.geograph.org.uk) website allows access to a geographical database, which is freely available to the public and which aims "to collect, publish, organise and preserve representative images and associated information for every square kilometre of Great Britain, Ireland, and the Isle of Man". This enabled me to locate many of the place-names mentioned in the literature.

Introduction

This series of short books on the place-names (not the actual places, historians please note!) of the old county of Northumberland, of which the present volume is the first, is intended to help general readers discover the possible origins — the etymologies — of the names of towns, villages, rivers and other natural features in our region. I chose to look at Northumberland before the 1974 local authority boundary changes because this reorganization is so recent compared to the history of most names, and it would impoverish the account to leave out the most important urban centre in the region, the city of Newcastle upon Tyne, and its environs. However, there is no denying the artificiality of even old Northumberland's borders as far as place-names are concerned, because our place-names have much in common with those north of the border in Scotland, as well as those in County Durham and the area now renamed Cumbria. So I make no apology for referring occasionally to the names of these adjacent areas as well.

The series began life as a single book, a revision of Godfrey Watson's *Northumberland Place Names. Goodwife Hot & others* (1970, Morpeth: Sandhill Press). However, Northumberland is a big county and it soon became clear that it would not be possible to produce a sensible treatment of the names without taxing the patience of publisher and readers alike. The material is presented as a series of journeys. In this way, I wanted to achieve my aim of seeing the names in their geographical context, as well as in the light of serious academic research. I hope that dividing the material into rough geographical areas will produce a readable account without too much duplication. The areas are: the western uplands of North Tynedale, Redesdale, Coquetdale and Glendale; the southern uplands of South Tynedale and Allendale; Tyneside and the Tyne Valley; and the Northumberland coastal plain between the hills and the sea. I have taken over Watson's narrative format, as this seemed to be a way of making the data more accessible while taking account of the context. This inevitably necessitates some repetition of material and I apologize for that in advance. All being well, the index will enable readers to find all the information they need quickly and easily.

The term "general reader" is not very precise. I imagine that some locals will be interested in the book, but it is aimed primarily at visitors to the region, and that is another reason why the text is arranged in the form of a series of tours of the region, mainly by car. That is also why some emphasis has been placed on pronunciation. English is renowned for its sometimes surprising spellings, and the domain of Northumberland place-names is no exception to this general rule. Where necessary, I have given a guide to pronunciation using the International Phonetic Alphabet (IPA). A guide to the IPA is supplied on Flap 2.

Sifting through the data has produced, almost as a by-product, a large database of Northumberland place-names based on the Ordnance Survey maps, so that the names found in these volumes correspond to the names on the map. Clearly, any serious place-name hunters will need to equip themselves with either the Landranger or Explorer Maps for the areas concerned (Landranger 74, 75, 79-81, 86-88; OL16, 42, and 43 will be most useful for this volume). Because of space constraints, every single name has not been discussed in detail, but the database provides comprehensive coverage of the area. However, I have attempted to explain some of the names other writers have avoided in their books, and these tend to be both the more difficult ones and the ones without any history, even though this can be fraught with danger (see below).

I have cited National Grid References for each place-name so each place can be found conveniently and unambiguously. Some writers have got away with discussing names which either turn out not to be in Northumberland at all (even the old county), or for which there is no real evidence — at any rate, I have been unable to find them on the map. Furthermore, it is no great discovery that there are sometimes several places with the same name, but it is surprising that they don't always have the same origin or meaning. For example, AYDON NU1115 probably means "hay hill", whereas AYDON NZ0066 means either "hay valley" or "island valley" and AYDON NY9256 comes from the Old English (OE) personal name *Ealdwine*. We can see immediately how National Grid References help to avoid confusion. In cases of doubt, the grid references refer to the location of the name on the map, not necessarily to the location of the actual feature. Obviously, it is impractical to give precise grid references for linear features such as rivers, streams and long ridges, and these should be deduced from the context.

It is possible to reveal distinct etymologies for place-names because scholars have, over many years, scoured manuscripts and other sources from the medieval and early modern periods, noted the old forms they contain and identified them with modern place-names. So their research reveals for instance, to use the example briefly cited above, that the early spellings of the *-don* element of the three places called AYDON probably reflects OE *dūn* "hill", or *denu* "valley", or is a development of part of a personal name (it can also reflect OE *tūn* "enclosure, yard"). Accordingly, I have used the three published sources which are generally reckoned to be the most reliable: Allen Mawer's *The Place-Names of Northumberland and Durham* and Eilert Ekwall's *The Concise Oxford Dictionary of English Place-Names*; finally, Victor Watts, *The Cambridge Dictionary of English Place-Names*. Ekwall's *English River Names* (Oxford: Clarendon Press, 1928) supplemented the brief entries in his dictionary, and other sources were used as necessary. Full details may be found in the section entitled "Further Reading" at the end of the book (p. 101).

Where old forms of a place-name exist, we are often able to give a plausible account of how the name developed, or at least to narrow down the possibilities. For the majority of place-names, however, we just have relatively modern forms, and our etymologies are much less secure. Sometimes, places with similar names in other parts of the country can provide a guide, but the example of AYDON has already shown that the modern form of names cannot be relied on absolutely, so the simple fact is that we can never be sure. Occasionally, the geographical and historical context can help us. Clearly, if the place concerned is on a hill, and we have to choose between an etymology indicating a hill (using OE *dūn* "hill") or a valley (using OE *denu* "dean, valley"), we should choose the former. Similarly, if we know that the place existed in Roman times, and the name is recognizably the same, it is pointless to force a later etymology on it. This is the case with YORK, which could easily be explained as "boar town" on the basis of Old English forms such as *Eferwic* (OE *eofor* "boar" + *wīc* "settlement") if a Latin source from AD 4 had not left us with *Eboracum*, which ties in with *Cair Ebrauc* in Nennius' *Historia Brittonum*, dated to ca AD 800, and so revealed it as a Brittonic name. The Welsh still call the city *Caer Efrog*! Conversely, if we know from historical sources that a settlement was not founded until the late medieval period, an earlier etymology would be illogical. Unfortunately, few cases are so clear-cut, and there is almost always room for doubt. Where historical sources do not provide an early name, I have tried to use comparison with other place-names to suggest a possible origin, but ultimately such speculation is just harmless fun. I can't stress too much that this is meant to be a popular rather than an academic book.

Nevertheless, a few words of warning to academic purists are probably still in order. First, it is common practice to say that the second half of a name like WELDON reflects Old English *denu* "valley", but this does not automatically imply that the name dates from the Old English period. It could date from the Middle or Modern English period and contain a later form, but it is more convenient to use the shorthand rather than list all the possible forms from every period in every case. Second, relying on published sources has its drawbacks, such as the danger of not recognizing other writers' assumptions, or worse still repeating their mistakes. Third, Northumberland is a big county and it would have been impossible to visit all the places concerned. Many householders and local experts may well have the key to solve a difficult etymology, so I would be most grateful for information and corrections from them.

Apart from Old and Middle English, other languages have also contributed to the place-names of Northumberland, such as Latin, Norman French, and Brittonic, the language of the indigenous population before the English arrived. Many people assume that Norsemen settled in Northumberland, but there are surprisingly few Scandinavian names in the county, and certainly no evidence to indicate Scandinavian settlement. What I mean is that the characteristic markers of Scandinavian settlement (such as names ending in *by*, like WHITBY, MICKLEBY, BARNBY, DANBY, and names ending in *thorp(e)* like UGTHORPE, MABLETHORPE, WRENTHORPE) are all absent in Northumberland, where OE *tūn* "enclosure" and OE *hām* "homestead" are to be found instead. There are no streams named using *beck* (Old Norse *bekkr* "stream"), as in the GLAISDALE BECK which runs through GLAISDALE, or the SANDSEND BECK which falls into the sea at SANDSEND, or the MICKLEBY BECK which runs through the valley between MICKLEBY and UGTHORPE. Instead, reflexes of OE *burna* predominate. If Northumberland had been settled by Scandinavians, we would expect to find a distribution of Scandinavian names similar to that found in Yorkshire, but we find nothing of the kind. Instead, there are a few possibles, but no pattern of naming which would indicate settlement. See the map in the illustrations section.

It is also surprising that, even though the place-names of the county are overwhelmingly English, there are many more possible Brittonic names in Northumberland than I ever thought to find. For example, apart from the ones other scholars have noted, such as TROUGHEND NY8591, OTTERCOPS NY9588 and YEAVERING NT9229, there are others such as BRAYDON NT8921, BURNT TOM NY6286, WALL NY9168, WIRCHET NY6297, possibly BLAEWEARIE NU0822, NU0639. There is also a series of place-names in *-don*, *-den* and *-ton* which are usually derived from OE *dūn* "hill". However, as these are hills on which the hill-forts are located, they could equally well be derived from Brittonic **dūnom* (as in *Camulodunum*, the name for the Iron-Age settlement at COLCHESTER). Fairly well-documented examples in this section of Northumberland include: FAWDON HILL NY8993, GROTTINGTON NY9769, LEIGHTON HILL NY9095, CRIGDON HILL NT8605, LOUNDON HILL NT9408, HORSDON NT9827, HUMBLETON HILL NT9628, FAWDON NU0315, SHAWDON HILL NU0813, and BURRADON NT9806. On the other hand, some postulated Brittonic etymologies, such as those for CARRYCOATS NY9279 and DUNTERLEY NY8283, have to be rejected, as will become clear later on.

It is impossible to write a book of this sort without having recourse to some technical terms. I have tried to keep these to a minimum, but cannot eliminate them entirely. A short glossary of technical terms used is given at the end of the book, on page 102.

Finally, I address the question of typographical conventions. First, small capitals are used to indicate place-names, and I have tried to list them as close to their grid references as possible. This also helps with disambiguation, e.g. AYDON NU1115 versus AYDON NZ0066. Second, old forms of the name are given in italics with a date, e.g. CARRYCOATS NY9279 (*Carricot* 1245), even though the precise date is often less important than the period in which the name was formed. It is also usual to italicize foreign words and words under discussion in the text. For example: the word *wood* comes from OE *wudu*. Third, because in place-name research we often deal with languages no longer spoken, and this also applies to Old English, it is sometimes necessary to reconstruct a word, such as **bōcen* "of beech" from the OE word *bōc* "beech" (this is formed in just the same way as Modern English *wooden* from *wood*), and these reconstructed forms are marked with an asterisk. This is to show that, although they have not been found in any actual texts, there is good reason to suppose that they existed in the language. Fourth, as in the last sentence but one, meanings are indicated by inverted commas. Fifth and last, I have used square brackets to indicate pronunciation (see Flap 2) and angle brackets to indicate spellings, e.g. "OE [k] can be spelt <c> or <k>".

CHAPTER ONE
1. From Carter Bar to the Ottercops

Like the reivers of old, some visitors to the western uplands approach from the north, but unlike the reivers they use the old trunk road (A68). For much of its course, the A68 follows DERE STREET, also sometimes known rather confusingly as WATLING STREET, the Roman road between Corbridge and the northern outposts of the Roman Empire in Britain. The first name, DERE STREET, probably comes from the name of the old kingdom of Deira (found in Bede as *Deira*, about AD 800, and in the Gododdin poem, ca. AD 600, as *Deivyr* and *Deor*), which could in turn derive from the plural of *dwfr* "water, river". Interestingly, the Gododdin poem is in early medieval Welsh and tells of the British tribe called the *Votadini* and their ill-fated expedition from their capital at EDINBURGH (*Eidyn*, about AD 600, *Edenburge* 1126) down the Roman road to meet the invading Saxons in battle at CATTERICK (*Katouraktonion*, about AD 150), which takes some unpacking if you're an Englishman who thinks that the ancient Britons spoke English. The truth is that they spoke a language we now call Brittonic, which later became Welsh, Cornish, and Breton. Some vestiges of the Brittonic spoken in Cumbria are preserved and usually called Cumbric, a term sometimes used to designate the language of the old north of Britain. The tribal name of the *Votadini* is probably preserved in the hill-name WODEN LAW NT7612 just over the border, where there are the remains of an Iron-Age hill-fort. The second name of the Roman road, WATLING STREET, is derived from OE *Wætlingastræt*, the first element being an early name for inhabitants of St. Albans in Hertfordshire, and is nowadays used almost exclusively of the Roman road between Richborough and Wroxeter.

Of CARTER BAR (*The Carter* 1695) NT6906, Tomlinson says that the name *Carter* comes from "the Celtic *cart*, a height or hill", but this word does not exist. There is Welsh *garth* "mountain ridge, promontory", common in place-names, which could be derived from Brittonic **garto-*, but there is precious little corroborating evidence for a possible change of initial [g] to [k]. The simplest explanation is from ME *cartere* "a driver of carts" for the first element of the name (and an association with CARTERHOUSE NT6707 over the border is obvious); the word *bar* in the meaning "toll-house gate, barrier" also dates back to ME. On the other hand, it could be a popular etymological development of OE *ceart* "rough common", in view of the terrain (as in BLACK CARTS NY8871). In the case of CARTER FELL, *fell* is probably not in the first meaning of a hill or mountain typical of the Lake District, but in the OED's meaning of a "wild, elevated stretch of waste or pasture land; a moorland ridge" which seems to occur chiefly in the north of England and parts of Scotland. CARTER PIKE NT6904 along the footpath to the south is a strange name, as one would expect it to designate a much more angular peak, but it probably refers to the cairn, as most Northumberland hills named *Pike* seem to have a cairn at their summit.

On the left-hand side of the road, the border fence crosses various high points. ARKS EDGE NT7107 appears to take its name from a farm just to the north (ARKS NT7108, which may reflect the use of the word *ark* to indicate "wooden bins for meal" and the like). The use of *edge* to mean "the crest of a hill" seems to be ME, although the word itself is older (OE *ecg*). LEAP HILL NT7207 may be so called because of its height, but it is not appreciably higher than the surrounding peaks and there is an area called HUNTER'S LOAP NT7206 just to the south and this may provide the motivation (Nthb. *loap* "leap"). FAIRWOOD FELL NT7307 may refer to the forested areas to the north, and we should note FAIRLOANS NT7508 "pleasant lanes" and FAWHOPE NT7409 "variegated valley" also just over the border. CATCLEUCH HILL NT7406 is named after the narrow steep-sided valley and the first element (OE *catte*) shows that wild cats once roamed the area. PHILIP'S CROSS NT7406 is a cairn on its western side, but there is no longer a cross and I have found no information on the man who gave his name to it. HUNGRY LAW NT7406 reflects either *hungry* in the sense "infertile, barren", or, in view of the patches of grazing, OE **anger* "green sward" as in ANGRYHAUGH NT9205). In GREYHOUND LAW NT7606 a hunting connection is not unlikely. The border fence joins the Pennine Way on the other side of a hill intriguingly named THE HEART'S TOE NT7606 (*heart* probably reflects OE *heor(o)t* "stag"). *Hill* is of course an English word, but *law* "hill" (OE *hlāw*) is also common in Northumberland and in other parts of the country too (*law* appears as *low* in the South and Midlands, e.g. MERRYTON LOW).

On the right-hand side of the A68, the ridge of Carter Fell rises up CATCLEUCH SHIN NT6806: *shin* is a Scots dialect word for "the sharp slope of a hill". As with THE HEART'S TOE mentioned above, it is very common to transfer names of parts of the body to topographical features. Another small valley called CATCLEUCH NT6806 falls away towards CARTERHOUSE NT6707 on the Scottish side. The ridge then climbs over a series of largely indistinct summits. KNOX KNOWE NT6502 may have been so called because of the early modern practice of naming hills after itinerant Presbyterian preachers (PADON or PEDEN HILL NY8192 is another possible example). Alternatively, we may be dealing with a reflex of Brittonic **knukko-* (Welsh *cnwch* "swelling, hump"). There are no old forms for DUNTAE EDGE NT6402, but the first element may contain Brittonic **dūnom* "fortified place" and be associated with DUN MOSS NT7007 on the other side of the border; the second element *-tae* may be the same as the *toe* in HEART'S TOE. THE TROUTING NT6502 may reflect MnE *trouting* which, apart from the activity itself, can mean a small stream in which trout are found, but we are at quite an elevation here. In HAGGIE KNOWE NT6301 (just over the border), TURFY KNOWE NY6097 and RUSHY KNOWE NY6198 (*Knowe* is pronounced [aʊ]), the epithets *turfy*, *rushy* and *haggie* (MnE *haggy* adj. "boggy and full of holes") describe the summit tops more closely, or perhaps how they were originally. RUSHY KNOWE, for example, is now covered with trees. In WYLIES CRAIGS NT6401, the second element is the Scots form of *crag* "rocky outcrop", and the first could just possibly be OE *wilig* "willow" as in WILEY SIKE NY6470. The magnificent KIELDER STONE NT6300 gives its name to the KIELDERSTONE CLEUGH NT6400, but takes its name from Kielder by the headwaters of the North Tyne. It has been suggested that PEEL FELL NY6299 is named after a pele tower, but there are no such buildings in the immediate vicinity and, given the location, it is more likely to be ME *pēl* "palisade, pole" or ME *pele* "a triangular shovel" because of its shape.

The river REDE (spelt *Rede* AD 1200) is the largest tributary of the North Tyne. The watershed of the river is also known as REDESWIRE. A *swire*, from OE *sweoru*, is a gentle depression between two hills; the site of the Redeswire Fray of 1517 is just over the border at NT7007. The river flows through REDESDALE (spelt *Redesdale* AD 1075) and falls into the North Tyne at REDESMOUTH NY8682 — there are no early forms of this name. Ekwall takes the name of the river to be OE *rēad* "red" and it reflects the local pronunciation of MnE *red* [ri:d]. However, the Rede does not strike me as running particularly red, but this may be because its flow is now restricted by the reservoir at Catcleugh. There are outflows of water coloured with red ochre here and there, but these are not unique to the Rede. On the other hand, Heslop lists *Reed-watter* meaning "chalybeate water" (i.e. that containing iron salts), so there is plenty of scope for popular etymology here and REDEWETTER is an old name for the Rede. An alternative explanation would be the Brittonic root **ret-* found in Welsh *rhedeg* "to run", and anyone who has seen the Rede in spate will appreciate the appropriateness of this name. It is also not beyond the bounds of possibility that REDESDALE acquired its name from the Reed family whose long-established seat was the farm at Troughend, but this would have to have occurred before 1075. The name does have a genitive *-s-* (unlike TWEEDDALE, TYNEDALE or COQUETDALE), and this *-s-* is often an indicator of personal names, but this may have been added because the name of the river was no longer well understood. There are many similar cases of popular etymology. If the name of the river is derived from a personal name, it may originally have had another name entirely, perhaps **Brēmia*, preserved in the name of the Roman fort at HIGH ROCHESTER, namely *Bremenium*. There is one final, perhaps rather convoluted, explanation for the name of the Rede. Modern place-names do not necessarily refer to the original settlements. So *Bremenium* becomes *High Rochester*, presumably in the early Middle Ages, as the evidence towards the end of this chapter suggests. This allowed people to call the village now on the main road Rochester, but this settlement did not exist in the early Middle Ages. The nearest settlement we can be sure existed at the time of the Roman occupation was the camp on the northern side of the Rede just south of the Redesdale Arms. This area is now called Horsley (see below), but as the name is English it must have been called something else beforehand. As the Roman road through Horsley and the Redesdale Arms crosses the Rede at a ford (there is no trace of a bridge as far as we are aware), protected by another camp on its southern bank, I suggest that the northern fort at least was called **ritu-castra* "ford-camp" (W *rhyd* "ford"), which could easily have developed to **Rede-chester* (subsequently re-formed as *Rochester*). The river itself would then be a back-formation from the original name of the Roman camp, and the name used for the nearby Roman settlement, modified perhaps by analogy with Rochester in Kent, or the name of the Reed family.

It is also impossible to say how old the names of the tributaries of the Rede above Catcleugh Reservoir or the valleys which contain them might be. BLACK CLEUGH NT6905 probably reflects the peaty soils, whereas CROSS CLEUGH NT6905 possibly indicates its position relative to the BATEINGHOPE BURN, as there are no wayside crosses in the immediate vicinity. The normal word for a stream in this area is *burn* from OE *burna*, as was mentioned in the introduction. The COOMSDON BURN and the CHATTLEHOPE BURN to the west of the A68 and the LUMSDON BURN to the north of LUMSDON LAW NT7205 as well as the RAMSHOPE BURN and SPITHOPE BURN on the eastern side all reflect this English word. There are no early examples for any of these names, so much guesswork is involved in teasing out what the first elements might mean. The original meaning of *hope* is uncertain. OE *hop* as in BATEINGHOPE seems to have meant either "enclosed land in a fen, part of a farm" or "valley", and the latter meaning is more suitable here. The first element of BATEINGHOPE may be MnE *baiting* in the sense of "hunting with dogs" (cf. GREYHOUND LAW). RAMSHOPE (*Rammeshope* ca. 1230, possibly the name of a medieval village) may contain either OE *hramsa* "wild garlic", OE *hræfnes* "ravens'", or *ramm(e)s* "ram's" or "rams": take your pick! It gives its name to the RAMSHOPE BURN which flows into Catcleugh Reservoir beside RAMSHOPE FARM NT7304, which was built in the mid-19th century. RAMSHOPE LODGE NT7204 is more modern (it was built in 1905). COOMSDON may reflect either OE *camb* "rocky ridge" or even OE *cumb* "narrow valley", an obvious early loan from Britt. **kumb-* "valley" (W *cwm* etc.). If it is a valley, the second element might be pleonastic *denu* "valley", which is not impossible in view of the presence of other Brittonic names in the area, but we should note that OE *camb* is rendered *Camb-* in CAMBO NZ0285. On the other hand, if it is a ridge, it may refer to the WHITE CRAGS NT6901 (a limestone outcrop, hence *white*) just to the south-west below which the burn rises, and the second element might then be OE *dūn* "hill" used pleonastically (*-don* in Northumberland may reflect *dūn* or *denu*). The existence of COMB HILL and THE COMB, which surely reflect OE *camb*, renders this plausible. The *-don* of LUMSDON LAW NT7205 is also very likely to be OE *dūn* "hill", and ME *lum* means "a well for the collection of water in a mine" (perhaps nearby TATE'S WELL NT7304 offers corroborative evidence for this etymology), although the word could possibly be earlier. There are Welsh adjectives *llwm* "bare, barren" and *llymes* "naked, bare", which would describe the bare hill accurately, and the latter offers an explanation for the *-s-*, but this could also derive from the compound underlying *llymfaes* "bare field, uncultivated land". WHITELEE NT7105 itself may contain *white* (OE *hwīt*) in the sense of "dry open ground" as opposed to *blæc* "woodland and black-land growing heath" and another form of the OE *hlāw* "hill", as in KIRKLEY NZ1576 (PNND 129), rather than OE *lēah* "clearing" which is the usual reflex of names in *-lee* and *-ley*.

The CATCLEUGH RESERVOIR, which now supplies water to Tyneside, takes its name from yet another CAT CLEUCH NT7403 (the slight difference in spelling is immaterial), the narrow valley which runs northwards from CATCLEUGH FARM NT7403, built in the late 18th or early 19th century; CATCLEUGH HOUSE NT7403, originally the home of the reservoir manager, dates from 1903. No evidence has so far been found to support the suggestion that there was a medieval village at Catcleugh, although much has been inundated with the building of the reservoir. Flooded places include BABSWOOD KIRK NT7402, "huge masses of disjointed and fallen rocks [...] where the covenanters, who fled from persecution in Scotland, held their assemblies" (Hodgson 2.1.135), and the farm BABSWOOD (probably "Babba's wood"; Babba may be the same lady who gave her name to BAMBURGH NU1834, spelt *Bebbanburh* in AD 547). Part of another farm, CHATTLEHOPE NT7302 (*Chetilhopp* ca 1320, *Shetilhop* 1317) still remains. This may mean "kettle-shaped hope" from OE *cietel* "kettle" or may simply contain the OE personal name *Cietel*. The name also lives on in CHATTLEHOPE CRAG NT7302 and CHATTLEHOPE BURN, which rises on the southern slopes of GIRDLE FELL NT7001. A little way to the south-west is the *Girdle Stone* from which GIRDLE FELL probably takes its name, a *girdle* being the local variant of *griddle*, the "circular plate of iron which is suspended over the fire and upon which cakes are baked or toasted". Just south of Girdle Fell, WOOL MEATH NY7099 and WOOLMEATH EDGE NY7199 naturally make one think of wool (OE *wull*), but the sheep ranches of Redesdale are a relatively recent phenomenon. Interestingly, OE *wulf* "wolf" regularly yields place-names in *Wool*-, such as WOOLAW NY8298 "wolf hill" further down the valley, so this was probably a landmark or other feature marking a boundary (ME *meathe, meith*) shaped like a wolf, or a place where wolves were to be found. Of the places above Chattlehope, CADGER BOG NT7301 probably contains the same first element as CATCHERSIDE NY9987 (spelt *Calcherside* 1270) and means "cold cheer" from ME *chere* in the older sense of "face, aspect", while ELLIS CRAG NT7401 may well contain the same personal name as ELISHAW (see below) and mean "Elli's crag". Of the *Jock* in JOCK'S CRAG NT7502 above Byrness not even such tenuous traces as these remain.

BYRNESS village NT7602 ['bɜːnes] (occasionally called THE BYRNESS locally) is situated in trees to the south of the A68, and was built by the Forestry Commission in the 1950s. The name itself is much older, meaning "burial mound on a headland" (OE *byrgen* and OE *næss*), but what could it refer to? There is a cairn on BYRNESS HILL NT7703, but this is not old, and it is possible that the prehistoric four-poster stone circle and round cairn (now called the THREE KINGS NT7700, as one of the uprights has fallen over) inspired the name, although the site should be imagined without the conifers. TOD LAW NT7700 contains the local name for a fox (ME *todde*) (there is another TOD LAW down the valley NY8397), as does TOD KNOWE NY7699. Just down the road from Byrness is THE RAW NT7601, which is usually explained as OE *rēw, rāw* "row" (of houses), and this name could then have been transferred to RAW HILL NT7700 just to the south. There was an Iron-Age village on Raw Hill, and this could have been the original motivation for the name, which appears in Hodgson as *Breadless-row*, alias *Breadless-straw*. This must be OE *bred* "board" plus *lēahas* "clearings" and the element *raw* (see the discussion of THE RAW

NY9498), but Hodgson's variants show the power of popular etymology, as *straw* is probably not original. LAD'S CLOUGH NT7600 on the other side of Raw Hill may contain the element *lade, lode* "an aqueduct or channel which carries the water to a mill" (Heslop 435).

COTTONSHOPEBURNFOOT NT7801 is said by some to be the longest place-name in England, but it is also one of the more transparent, being at the lower end of the COTTONSHOPE BURN, which flows into the Rede near this point. The farm at COTTONSHOPE NT7904 (spelt *Cotteneshopp* ca 1230) is named after the original settler, who was perhaps called Cotta or Coten, and his valley or enclosure. COTTONSHOPE HEAD NT8006 is the farm at the top end of the valley just off the road which joins the course of Dere Street. Overlooking the valley are DOUR HILL NT7902, GREAT DOUR NT7903, BLACKKIP NT7904 (compare KIP HILL NZ0267 and KIP LAW NY7150, which may contain a form of ME *kep* "keep") and LOAN EDGE NT8005, which might mean the "edge of the lane" (Nthb. *loan*), perhaps referring to a drove road, or a sheltered place where cows were gathered to be milked (Heslop 454). The *Dour* names may be the same as MnE *dour* "fierce, sullen", but this would be a late borrowing via Scots from Latin *durus* "hard". Preferable would be OE *dor* "door" which is sometimes found in place-names in the sense of a mountain pass (as in DORE in Derbyshire); alternatively, the first element of DORCHESTER is Britt. **durno-* "fist" (Welsh *dwrn*) and either "pass" or "fist" would provide a satisfactory meaning here.

BLAKEHOPEBURNHAUGH NT7800, the flat alluvial land (OE *halh*) where the BLAKEHOPE BURN flows into the Rede, runs the longest name a close second. The *Blake-* element could be MnE *black* (= OE *blæc*) after the colour of the soil or coal outcrops or, rather confusingly, OE *blāc* "pale" after the colour of the rough grass. A number of Northumberland place-names present us with this problem. Whatever the origin, BLACKBLAKEHOPE NY7599, once an old farmhouse but now a picnic site, acquired one or the other element again, probably in an attempt to clarify the meaning. The BLAKEHOPE BURN rises near BLAKEHOPE NICK NY7198, the highest point on the Forest Drive between Redesdale and North Tynedale, *nick* being a common local place-name element meaning a "gap in the hills".

On the main A68, drivers find themselves climbing through the last of the forest below MALLY'S CRAG NT7900 (which may possibly contain the Christian name *Mary* as first element) to emerge into the open landscape of the MOD ranges on the left and the treeless hills on the right. BELLSHIEL [bɛl'ʃi:l] NY8199, the farm on the right, is probably not the original location of the name. The second element (ME *schēle* "hut", or "summer hut") refers to the common practice of transhumance in the area, but the first element *Bell-* is more problematic and has implications for many place-names. OE *belle* "bell", could be transferred to bell-shaped objects and therefore mean a hill (as in YEAVERING BELL NT9229) or a bell-shaped pit and indeed the ruins of BELLSHIEL PIT COTTAGE beside the road to BELLSHIEL LAW NT8101 provide evidence of relatively recent mining activity in the area. There are over eleven bell pits and six waste heaps nearby.

The BLACK BURN, which runs into the SILLS BURN NT8202, may have been named after coal measures in the vicinity. Over the river from Bellshiel Farm is BURDHOPE ['bɜːdəp] NY8198, sometimes spelt as *Birdhope*, "bird valley", but this is not the form on the map (also spelt *Burdop* 1595). Perhaps this name was originally applied to the old settlement on the lower slopes of BLACKWOOL LAW NY8098 "black wolf hill". WOOLAW FARM NY8298 "wolf hill farm" is just downriver.

On the opposite bank of the Rede, the site of the old REDESDALE CAMP NY8299 (not an ancient British camp but part of the Otterburn Ranges) lies on top of a rocky outcrop, marked BELLSHIEL CRAG NY8199 and also BURDHOPECRAG on the most modern map. This gave its name to BIRDHOPE CRAIG HALL, burned down in 1957. Hodgson (1827, 2, 1:147) says: "Birdhope Crag is a hamlet, consisting of a good dwelling house, a presbyterian meeting house, and a few cottages: seated on a fine knoll on the right bank of the Rede, next to which its side is steep and rocky, and clad with a birchenholt, and on the south glides swiftly down into the haughs that hem the willowy borders of the river. Travellers, who have resided in the neighbourhood of Rome, have remarked on the striking resemblance between this place and Fidenae." He seems to be referring to what is now known as ROCHESTER NY8298. One thing is certain: the name BRIGANTIUM NY8398 for the now mothballed archaeological reconstruction centre is new, coined by the owner on account of the British tribe the *Brigantes*, whose centre of power was probably further south. (It is a genuine name, however — the antecedent of *Bregenz* in Austria, for instance!) The modern lane to HIGH ROCHESTER NY8398 (*Roff'* 1208, *Rucestr* 1242, *Rouschestre* 1325), doubles back on the main road and runs almost due north. If not named after ROCHESTER in Kent (this may be the explanation of the earliest form *Roff'*, perhaps an abbreviated form of *Roffensis*), it may contain OE *hrōc* "rook" and mean "the fort where rooks nested" (but see the discussion of the name of the River Rede above). Indeed, corvids gave their name to nearby CORBY PIKE NT8401 and RAVENS KNOWE NT7706. Another possible explanation is ME *rouk* "mist, fog" as in ROOKEN EDGE. The earliest name of this northerly outpost is recorded as BREMENIUM, which possibly preserves the original name of the River Rede, shared by the BREFI in Cardiganshire. The change of [m] to [v] is quite regular and the name is possibly recorded later by Nennius in *Historia Brittonum* "The History of the Britons" 56 as *Breguoin* and *Breuoin*. The farm of DYKEHEAD NY8398 is on the course of another Roman road which strikes out towards Coquetdale. A *dike* is a local word meaning "fence, hedge, ditch, earthen or stone wall" and the farm is located at the top or head, so this is logical enough. HILLOCK NY8399 "little hill" is another transparent name roughly on the same contour to the north. HUEL KIRK NY8399 is marked just to the north of HUEL CRAG NY8398 on the map. *Kirk* may be a reference to its possible use as a Presbyterian meeting place (like Babswood Kirk) before the chapel at Birdhopecraig was built (i.e. the one near the original BIRDHOPECRAIG HALL dated 1682; the one in the village was actually built in 1826). Alternatively, it may derive from Britt. **krūko-* "hill", as in KIRKLEY NZ1576. The first element *huel* is likely a compound of *heugh* [hju:f] "precipitous hill, cliff" (as in CALLERHUES CRAGS NY8486) and *hill*, so the name HUEL CRAG could mean "crag hill crag".

Looking towards the river from Rochester there is a small wooded knoll called TOD LAW NY8397 "fox hill". The WIND BURN — pronounced ['waɪnd] and it is indeed a very bendy stream — flows into the Rede just behind, rising in an area of forest which masks LOAF HILL NY7698 (probably named after its shape, as the word *loaf* can be transferred to hills). Also hidden from view are HINDHOPE LAW NY7697 ['haɪndhəʊp] (either OE *hind* "female deer" or the local word *hind* "raspberry"), WETHER LAW NY7695 (probably OE, ME *weðer* "castrated ram") and ROOKEN EDGE NY7895 ['rʊkən] (ROOKEN KNOWE NY8096 is marked to the east). This last could contain OE *hrōc* "rook", but is perhaps more likely to reflect *rouk* "a mist, fog, and the drizzle accompanying it" (Heslop 584). The BLACK HILL NY8195 overlooks STRUTHER BOG NY8195 (*struther / strother* "a marsh or swamp where rushes grow" Heslop 704). On the other side of a small valley is EVISTONES NY8396 (a village called *Cleughbrae alias Evington* in 1618). This is one of the many deserted villages of Northumberland, in this case since the end of the 17th century, and the dilapidated buildings are a likely motivation for the remodelling of the second element of the name, i.e. *-stones*. Originally, it probably meant "the enclosure (OE *tūn*) of Geofa's or Eofa's people" (this by comparison with the early forms of EVINGTON in Leicestershire). While on the matter of personal names, KELLY'S PIKE NY8195 and KELLYBURN HILL NY8395 are unlikely to have anything to do with recent Irish immigrants to the area, although there were plenty engaged in the construction of the reservoirs, railways and roads. Kelly possibly derives from the *keely-haak* "kestrel [...] the commonest falcon in the North of England" (Heslop 416). CLEUGHBRAE NY8396 on the burn itself reflects ME *clogh* "ravine" and ME *brae* "slope", but there are no reliable early forms. The farm ASHTREES NY8395 was built around 1840, and is also a modern name. Below Evistones is NETHER HOUSES NY8397 which likewise dates from the early 19th century, although there are two Romano-British settlements to the west and a number of other sites from that period which suggest that a substantial British settlement had a clear view over the valley towards the Roman fort at High Rochester / Bremenium.

On what might be termed the Roman side of the river is STOBBS (farm) NY8397, which is just the plural of OE *stob* "(tree)stump" and was a common name for cleared land. Another word for a clearing was OE *lēah* which we find in HORSLEY NY8496 (*Horselahe* 1604), meaning "horse clearing", not an uncommon name. In this case, the name may well originally have referred to what is now the Redesdale Arms, an old coaching inn once known as the Horsley Inn. It is on the course of Dere Street with a Roman camp just to the south, and one might imagine it to have had its origins in the *vīcus* or native settlement outside the fort. Dere Street crosses the modern A68 south of the inn and runs past BAGRAW (farm) NY8596 and BIRKHILL NY8595 (*birk* is the local form of MnE *birch*) before fording the river and following the course of the

A68 near BLAKEHOPE NY8594 (we will return to this route). BAGRAW shares its name with a number of places in Northumberland and, like RATTENRAW NY8495 on the other side of the river, clearly contains the *raw* element discussed above (either *row* in the sense of "ramshackle building" or a word meaning "hill"). But what of *bag*? There are no early forms for the Redesdale BAGRAW, but on the evidence of others it may just possibly be a personal name or an early form of *badger* "itinerant seller, hawker". Perhaps it simply meant "badger hillside" (there are still real badgers in the area).

Opposite Birkhill is BENNETTSFIELD (farm) NY8595 which may contain a personal name *Bennett* but more likely reflects OE *beonet* "bent, i.e. a species of grass". We have the same element in BENNETT'S RIGG NY8596 to the north. The turnoff to the left is marked YATESFIELD ROAD on the map and leads over YATESFIELD HILL NY8597 to STEWART SHIELS NY8698 and the course of the spur of the Roman road which connects High Rochester with Coquetdale. The 7th-century Saint Eata, Bishop of Hexham and Lindisfarne, would have travelled this way to his interests at Melrose, as there was no other road apart from Dere Street, and it is possible that he gave his name to YATESFIELD as well although there is no direct evidence. YATESFIELD (farm) NY8697 is mid 19th-century. STEWART SHIELS probably meant the summer huts of a steward (OE and ME *stiward*, compare STEWARD SHIELD MEADOW in Co. Durham NY9843, but the forms with <t> are especially characteristic of northern ME).

The A68 crosses the Rede at ELISHAW NY8695 (*Illescagh, Illeschawe* 1278), perhaps the site of a medieval hospital or hospice. However, the name refers to the copse of a person called something like the Eli of the Domesday Book (but we cannot say for certain, as the Domesday Book does not cover Northumberland). The A696 carries straight on along the Rede and dips to cross the DURTREES BURN, to the east of which POTTS DURTREES (farm) NY8797 is situated. The early forms are represented by *Dortrees* (1275) and the name Potts, still common in the area, had been added by the 17th century. *Dortrees* probably means "doorposts". SHITTLEHEUGH NY8694, either "Scytel's haugh land or shuttle-shaped haugh" is near the road by the river, but there is an earlier bastle and a Romano-British settlement on the hill to the south. If BLAKEMAN'S LAW NY8795 is considered together with BLACKMAN'S LAW NY7498, and perhaps BLAKELAW NZ2166, the usual derivation from a personal name such as *Blakeman* becomes less likely. These names could contain OE *blæc* "black" on account of the coal outcrops in the area, but could equally well reflect OE *blāc* "pale, wan, pallid", because of the poor grassland. The -*man*- element remains a mystery but may be a pleonastic element from Britt. *moniyo- (W *mynydd* "mountain", as found in place-names such as LONGMYND and MINTON in Shropshire, MINDRUM NT8432 etc.). GREENCHESTERS NY8794 is the next farm on the left, reflecting OE *grēne* "green" and OE *ceaster* "Roman station" (from Latin *castra* "camp"), maybe referring to the Romano-British farmstead which encloses the remains of a medieval building just to the north. It is worth noting that the plural *chesters* is a peculiarly Northumberland element. Driving past the site which commemorates the Battle of Otterburn fought in 1388, and the inappropriately named Percy's Cross (as it it here that Earl Douglas is said to have fallen), OTTERBURN village NY8893 itself (*Oterburna* 1158, i.e. OE *oter* "otter" + OE *burna* "stream") is located in a small depression north of the Rede. The road to Otterburn Camp and Otterburn Hall, built in 1870 as a country house for Lord James Douglas, follows the course of the Otter Burn towards DAVYSHIEL COMMON NY8997 (*Davisel, Daveschole* 1344), which contains a personal name like *Davy* as first element. The burn rises on the slopes of GREYMARE RIGG NY8998. There are a surprising number of "Greymare" names in Northumberland, probably indicating boundary stones. The intriguingly named FIDDLER'S WOOD NY8898 lies to the west (compare FIDDLER'S GREEN on Holy Island; FIDDLER'S HAMLET in Essex). The name may be a popular etymology, perhaps based on OE *fifealde* "butterfly", or refer to grain fiddles. BLACK STITCHEL NY9098 to the east seems to contain a second element OE *hyll* "hill" and the first might be OE *stūt* "gnat" as in STITCHCOMBE in Wiltshire, or OE *stig-stede* "place with a pigsty". GREENWOOD LAW NY8999 is to the north, but any green woods are long gone.

Before leaving the village, there is a turn to the right, which leads past Otterburn Mill, once famous for its woollen blankets, and over the river towards the A68 (Dere Street). WOODHILL NY8892 is on the right and has a transparent modern-looking name. OLD TOWN farmhouse NY8891 appears to date back at least to the 18th century, so perhaps it was named after the earlier Romano-British settlements nearby. BROWNCHESTERS NY8892, on the other hand, is a new name despite its appearance: this farm used to be called THE BOGG. On the other side of the road, a track now leads to GIRSONFIELD NY8993. The genitive ending *-(e)s* in the early form *Grenesonesfeld* (1331) points to a personal name, i.e. a field (OE *feld*) belonging to someone with a name like *Greenson*. Perhaps the modern form was influenced by the local form *girss, gyrss* "grass". Above that is FAWDON HILL NY8993 with its Iron-Age fort (certainly OE *dūn* "hill", perhaps from Britt. **dūnom* "fort, oppidum"). The first element could refer either to the variety of colours in the vegetation or the likelihood that these settlements were painted. HEATHERWICK NY8992 may contain MnE *heather* (and this has probably produced the name HEATHERY HILL NY8993 on the other side of the road), but another possibility is OE *hēahdēor* "deer". The second element *-wick* is derived ultimately from *vīcus*, the civilian settlement outside a Roman fort, which came to mean "farm" and was subsequently generalised to mean "farm" even where there was no Roman fort.

We do not take the road to Elsdon, but keep right and negotiate the sharp left-hand bend by MONKRIDGE HALL NY9092 which takes its name from a settlement referred to as *Munkerich* (ca 1250), reflecting OE *munuc* "monk" and OE *hrycg* "ridge", although it is unclear which monks were being referred to, or where the original ridge was. The settlements at HEATHERWICK, or MONKRIDGE (farm) NY9191 across the valley of the Elsdon Burn may be meant. The road crosses the burn and then, with HAINING HEAD NY9292 (ME *haining* signifies "enclosed land") on the left, and WETHER HILL NY9290 on the right (where the castrated male sheep, OE and ME *weðer*, were kept), the road climbs up to RAYLEES NY9291 (*Raleys* 1377) which means "roe-deer clearings" (OE *rā, rāha*). There is another turn to Elsdon at Raylees, where the RAYLEES BURN crosses the road which winds up past RAVENSCLEUGH NY9391 (probably "ravens' ravine") and RAYLEES COMMON, now under trees. BLAXTER COTTAGES NY9390 reflect the name of BLAXTER LOUGH NY9389 (the local spelling of Scots and Northern *loch*) and the now partly mothballed BLAXTER QUARRY, which used to supply stone for Edinburgh New Town. The name *Blaxter* is far from straightforward: the first element may be *bleak* "to make white or pale by exposure to light" (= MnE *bleach*) to which has been added the ending *-ster* seen in *baxter* "baker" and *webster* "weaver". Hodgson's explanation, that the name is derived from a "tarn or lough of *black water* in a peat moss [...] for *stur* and *stour*, in old language, meant the same as a water or a river" (p. 105) founders on the identification of the language concerned. The name of the River Stour is an unlikely connection, as this is generally derived from the Germanic stem found in ON *stór* "great", etc. To sum up, the word *blaxter* seems to have meant "bleacher".

Finally, we reach the masts on MOUNT GILBERT NY9489 near OTTERCOPS MOSS NY9589 on the left. Gilbert was the name of several members of the Umfreville family, who owned the land until the mid-12th century, but there is no direct evidence of a motivation. The name OTTERCOPS naturally makes us think of Otterburn, but the old forms (e.g. *Altercopes* 1265) show that a connection is unlikely. A more defensible explanation is Britt. **altā* found in Welsh *allt* "hill" and several place-names such as Y WENALLT "white hill" in Merionethshire, but also ALT near Manchester, plus OE *copp* "summit", and we would therefore have a word which had become unfamiliar being reinforced, pleonastically as it is called in the trade, by a more familiar word. Such cases are very common. There is also a Welsh word *copa* "summit", but this is generally treated as a loan from English. *Moss* is a northern word for a bog (also ME *mos*), which fits the boggy terrain very well. OTTERCOPS FARM and OTTERCOPS BRIDGE NY9588 are located nearby, and are secondary names.

CHAPTER TWO
2. From Elishaw to Stagshaw

Let us now retrace our steps to the point where the A68 crosses the Rede at Elishaw and skirts round a Roman camp on the left beside the modern farm of BLAKEHOPE NY8594, and follow the course of Dere Street for a while. Blakehope Fell is the area on the right and RATTENRAW (farm) NY8595 is back in the direction of Upper Redesdale. There was a Romano-British farmstead nearby and the conventional explanation is that RATTENRAW means "rat-infested dwelling(s)", from ME *raten* "rat", a loan from Anglo-Norman, plus ME *raw* "line of houses" as discussed above. We should note its proximity to BAGRAW NY8596 across the river, and that the name RATTENRAW is not uncommon in England — I have found nine so far. On the high ground behind is PADON HILL NY8192, which possibly recalls the name of the Scottish covenanter Alexander Peden (1626-1686), but the spelling does not fit. It may equally plausibly be yet another possible location for the 6th-century Battle of Mons Badonicus mentioned by the British monk Gildas, as the hill strategically overlooks Dere Street. To the north-east lies an area called GORLESS NY8293/NY8393. The second element of GORLESS probably means "clearings" (as in CATLESS NY8375, and the first could then be OE *gor* "mud" as in GORTON in Lancashire (now East Manchester).

DARGUES NY8693 is the location of yet another Roman camp but probably reflects a family name, deriving from OE *dæg(e)weorc* "day's work, daytime". The original farm was seemingly split between the Dargues and the Dunns (hence DUNNS HOUSES NY8692 further along the road), but the former have given their name to DARGUES BURN and a smallholding now called DARGUES HOPE NY8493 after one of the little valleys (OE *hop*), or OE *hop* may later have come to signify "portion of a farm". Before you reach Dunns Houses, there is a minor road to the right leading to TILESHEDS NY8592 and PIT HOUSES NY8191. Both are probably modern names. The OED has no entry for *tilesheds*, but the place-name appears elsewhere in Northumberland. The tiles in question were probably not used for roofing but for land drainage. GARRETSHIELS NY8693 (*Gerrardscheles* 1291) means "the summer huts of Gerard" (see DAVYSHIEL above) and has nothing to do with fortifications or upper rooms, as suggested elsewhere.

The next farm on the western side of the A68 is TROUGHEND NY8591, a name also found in TROUGHEND COMMON and as an old ward of Elsdon parish. The early forms (*Trocquen* 1242, *Trequenne* 1279, *Trehquen, Troghwen* 1293), suggest that it might be derived from Britt. **trebā-windā* "white farm", so the name would have been applied not to the old house (allegedly blown up by the owner to avoid "listed building" status) or to the pele tower which preceded it, but to one of the Romano-British farmsteads in the immediate vicinity. Turning right at the cross-roads signposted Bellingham, we cross the TOFTS BURN at TOFTS BURN BRIDGE NY8791. The name comes from TOFTS NY8592, which contains the common place-name element *toft* "a green knoll, especially one suitable for building". The TOFTS BURN rises on WHITLEY PIKE NY8291 on the course of the Pennine Way, which is marked by a cairn (this use of *pike* is very common in Northumberland).

Whitley is likely to be motivated either by the rough grass or the all-too-frequent snow and either a clearing or a hill. LORD'S SHAW NY8291 just to the south no longer has a copse and LOUGH SHAW NY8489 has no lake either, but these are sometimes transitory features. GREY STONE NY8290 and GREAT MOOR NY8590 seem to mean what they say. WETSHAW HOPE NY8589 is either the "valley or piece of land near the wet copse" and WETSHAW SIKE NY8689 is the "watercourse" nearby. DEER PLAY NY8490 is probably a place where rutting deer were to be found, as sites of this type are often called *play* (see for example COCKPLAY NY8872).

These hills are to the north of the B6320 which passes HARESHAW HEAD NY8588 on the right, said to have been the Hareshaw Colliery manager's house, and then proceeds over what is left of HARESHAW village NY8488. The name seems to mean "hare copse" (from OE *hara* "hare") rather than "grey copse" (from OE *hār* "grey, hoary"), and indeed hares can still be seen along this road, but no copses. The 19th-century farm HARESHAW HOUSE NY8487 lies between ABBEY RIGG NY8487 (but it is unclear to which abbey it belonged) and the HARESHAW BURN which used to supply the Hareshaw ironworks in Bellingham, but now provides a wooded walk from the old foundry yard to HARESHAW LINN NY8485. The word *linn* is a common local word for a waterfall, and there is an OE word *hlynn* "torrent" which is unique to Northumbrian and appears to be without cognates unless it is related to OE *hlynnan* "to resound". More probably, it originally signified the pool underneath, in which case it is borrowed from Britt. **lindo*- which we find in Welsh *llyn* "lake" (as well as *Lindum*, the old name for LINCOLN). We complete our excursion to Bellingham past SHEEL LAW NY8384 "summer hut hill" on the right on the road and HAININGRIGG NY8484 "enclosed ridge" (ME *haining* "enclosure") on the left. Before you reach the village of Bellingham, there is a turn to the right to GREENHEAD NY8383, one of several places with that name, and further along the track REENES (farm) NY8284 which is the plural of MnE *rean* "a deep furrow used for conducting drainage water from a field or other piece of ground, a water-furrow".

We now resume our journey on the A68, and on the other side of the B6320 junction, we cross the MILLER BURN NY8791. This name probably reflects ME *miller*, but it is unclear now where the mill was situated. The road climbs COCK RIDGE NY8790 (OE *cocc* "wild bird"), before crossing the BRIGG BURN, which probably gets its name from THE BRIGG NY8989, a farm by a bridge on the old drove road. At the top of the hill is another DYKE HEAD NY8889 and CORSENSIDE with the 12th-century church of St Cuthbert NY8989. CORSENSIDE is the name of the ecclesiastical parish and also appears in CORSENSIDE COMMON NY8688. It is spelt *Crossinset* in 1254. The development from this medieval spelling to the modern form is achieved on the one hand by metathesis of the first part of the name (*Cross-* becomes *Cors-*) and on the other by the replacement of what is probably the ending, either OE *(ge)set* "dwelling, fold" or *sǣte* "house", by the common element *-side* (from ME *sīde* "hillside"). The *-in-* bit becomes *-en-* because it is unaccented, and may belong with *Cors-*. One theory is that *Corsen-* is derived from an Irish personal name Crossán, and the influence of Celtic christianity on Northumberland is indisputable, but there is no clear evidence of a person called Crossán. However, as *Corsenside* is an ecclesiastical name, a connection with a cross of some description is likely (there is a medieval cross base in the churchyard). However, the word *cross* is not recorded in English until the 13th century, and is known earlier only from place-names, often with accompanying Scandinavian elements which is why some scholars have emphasized the Old Norse connection. The ultimate source of the word in all the languages concerned is Latin *crux* "cross", and it is important to consider that Welsh has borrowed both *croes* and *crwys* from the Latin, so an early Brittonic loan is not out of the question. We have already shown that there are other Brittonic place-names in the area, whereas Norse or Irish origin (ON *kross*, OIr. *cros*) would be unusual for Northumberland. Whether by popular etymology or not, we should note that the ground rises up westwards to CROSS LAW NY8689, and there is also a medieval wayside cross alongside the old trackway at WOODBURNHILL (farm) NY9186. The *-in-* element could simply be the preposition *in* "in, on" and the whole name *Corsenside* taken to mean "cross on the hillside" referring either to one of these other crosses or the church itself.

At CROW BRIDGE NY8888, the A68 crosses the CROW BURN in the dip (CROW WELL — a petrifying well — is also marked, all reflecting either OE *crāwe* "crow" or OE *crōh* "valley", just possibly OE *croh* "saffron") and then rises to COLDTOWN NY8988 (*Caldton* 1331) with its bastle house. It then diverges from the course of Dere Street at WOODHOUSE NY8887 (*Woodhouse* 1865) and runs down to WEST WOODBURN NY8986. Early forms are *Wodeburn* (1265) and *Wodeburge* (1287) so it means either "stream in the wood" or "fort in the wood". There is no stream with this name, but the LISLES BURN which rises on RAY FELL NY9585 and falls into the Rede at EAST WOODBURN NY9086 may be meant. The fort may refer to the Roman fort of HABITANCUM NY8986 to the south west of the main road, marked as RISINGHAM on the 1865 OS map. *Habitancum* is normally classified as Latin, although we might compare *Avitācum*, the Gaulish name of AYDAT in the Auvergne which might suggest Brittonic origin. *Risingham* ['rɪzɪŋəm] has all the hallmarks of an OE formation, "the village of Hrisa" like RISBOROUGH in Buckinghamshire, and the *-ing-* element of *Risingham* probably reflects an old genitive ending in *-n* rather than the "people of" marker in *-inga-* (or else we might expect the pronunciation to be ['rɪzɪndʒəm], as in Bellingham etc.). Alternatively, we may have a reflex of OE *hrīs* "brushwood" in *Risingham*, as in the nearby RISEY BURN which runs past SUMMIT COTTAGES NY9384, one of the the highest points on the Wanney / Wannie railway line (both spellings occur), which opened in 1865 and closed in 1966.

An excursion up this side valley takes you over the Lisles Burn at CHAIRFORD BRIDGE NY9086, a 19th-century house doubtless named after the bridge, itself probably on the site of a ford if the name is anything to go by, but the first element of the name remains obscure (possibly OE *cyr-* "church-"). There is no need to worry about HOT

HEADS NY9086: this is another local word for (the top end of) a wood (OE *holt* "copse, wood") with vocalization of the [l]. If you follow the road, DARNEY CRAG NY9187 ['da:nə] is on your right with DARNEY HALL NY9187, a local name, improbable as it may sound, given to "a strong cavernous fissure", so named according to Hodgson (p. 167) "probably from the water which in some parts of the fissure can be ascertained to be in it, by the splashing noise made at its bottom by throwing stones into it". The element *darney* could reflect OE *dierne, dern* "secret" as in DARNCROOK near Gateshead "a secret crook of land". DARNEY QUARRY NY9188 bears witness to this valley as a source of stone and the place-names are also interesting in this regard, e.g. WHETSTONE HOUSE NY9286, or GRINDSTONE SIKE NY9187, which comes down from STANIEL HEUGH NY9187 (OE *stānegella* "kestrel" literally "stone yeller"). Above THE DOD NY9187, a common name associated with a bare-topped hill, is HARTSIDE NY9287 with its cairn, the name reflecting OE *heorut* "stag" with WILLY CRAG NY9288 (possibly OE *wilig* "willow", as in WISHAW below), THISTLE CRAG NY9288 and TOD CRAG NY9387 (either *tod* "fox" or the same word as *Dod*) to the east. HIGH NICK NY9387 "high pass" has a sandstone quarry and further to the east again is WISHAW NY9487 with WISHAW PLANTATION and WISHAW PIKE to the north. These may reflect OE *wīþig* or *wilig* "withy or willow" or OE *hwīt* "white" if the name of the area called WHITE MEADOWS NY9487 is old and not itself derived from OE *wīþig*.

Three unnamed streams flow down from RAY FELL NY9585, GOWK HILL NY9585 (*gowk* is a local word for "cuckoo"), and CRANESTER BOG NY9586 (perhaps OE *cran* "crane" with OE *ēstre* "sheep-fold"), including one in CUDDY'S CLEUGH NY9487 (probably "Cuthbert's ravine"). They converge at THREEBURN MOUTHS NY9486, flow into two small lakes in LAKE WOOD (these names seem to be modern) and emerge over a weir as the LISLES BURN. This watercourse is probably named after the post-Conquest de Lisle family whose seat was at HALLYARDS NY9086. Only the fishponds remain near the old vicarage which is now called MIDDLE WOODBURN HOUSE NY9086. There is also a waterfall known as THE LINN and a house just to the north known as LINNHEADS NY9386. The Lisles Burn then flows through a wooded valley known as the LISLES BURN DENE (this name is modern, despite OE *denu* "valley"), past HAREWALLS bastle NY9628 (probably OE *hār* "grey" plus the common second element *walls*, which is often associated with delapidated buildings) and yet another BLAKELAW NY9286. This last is probably "black hill" in view of the coal workings. STIDDLEHILL COLLIERY NY9185 was worked commercially in the late 19th and early 20th centuries, but there is evidence of earlier drift mining. STIDDLEHILL (house) and STIDDLEHILL COMMON also retain the name, which is probably OE *stede* "place, site of a building" plus OE *hyll* "hill" or *hlāw* "hill" with modern *hill* added pleonastically (i.e. when the *-le* of *Stiddle* was no longer understood). COLD LAW NY9285, of which there are at least five in Northumberland, is just to the east. This one overlooks MIDGE HOLE NY9385, an apt name for most places plagued by the local tiger midges, and SONSY NOOK NY9385, which contains the Northumbrian word *sonsy* "lucky, good-looking, jolly, pleasant", so a "pleasant corner" despite the midges (or the name may reflect a touch of irony). RAY TONGUE NY9386 may be the tongue of land (OE *tunge*) or the fork of a river (OE *tang / *twang*) belonging to the township of Ray or where roe deer were found (OE *rā, rāha*).

Returning to the main road with the course of the Roman road on our right and climbing up the bank to the south of what is now West Woodburn, BROOMHILL NY9086 with a seemingly transparent name (i.e. the plant broom, OE *brōm*) is on the left by a road which leads to HIGH SHAW NY9185 (a common name). PARK HEAD NY9085 is further up the slope on the right near where the replica of Rob of Risingham was erected by the Redesdale Society in 1983, and in the valley below is CHESTERHOPE NY8985 (*Chestrehop* 1298), the Roman fort in question being HABITANCUM (see above). CHESTERHOPE COMMON is the name given to the moorland overlooking Ridsdale. Also overlooking the Rede as it winds its way towards Redesmouth are CRAGG FARM NY8885 and HINDHAUGH NY8784, the latter not indicating a crag (OE *hōh*, i.e. Nthb. *heugh*) but the flat alluvial land below (OE *healh*, i.e. Nthb. *haugh*). It is less easy to decide whether female deer (OE *hind*) or hindberries (local *hindberries* correspond to MnE *raspberries*) are meant. The road runs straight towards SARELAW COTTAGE which takes its name from SARELAW CRAG NY9185 above. The first element is most likely OE *sēar* "barren", so the name means "rocky outcrop on a barren hill" which describes it nicely. The road swings round to the right and crosses the BROOMHOPE BURN, which flows into the Rede below BROOMHOPE MILL NY8783. BROOMHOPE itself NY8883 (*Bromhop* ca 1250) is on the higher ground to the north, but named after the valley where broom grew. THE STEEL NY8982 was originally just another farm on a ridge (OE *stigol*) near the Broomhope Burn, but became associated with the Ridsdale ironworks as a testing site for guns (hence *The Gun Inn* in Ridsdale). Visitors often remark on the medieval castle just to the north of the village of RIDSDALE NY9084, but in fact these are the ruins of the engine house for the ironworks and the whole village was built in the mid-19th century. The name appears to reflect the local pronunciation of the valley below (i.e. *Redesdale* ['rɪdzdəl]), or some approximation to it and is of no antiquity at all. Indeed, *Redesdale* is the spelling of *Ridsdale* on most 19th-century maps.

On the way down from Chesterhope Common, FOURLAWS (farm) NY9082 ("four hills" although it is difficult to say precisely which hills were meant) and the entry to THE STEEL are on the right, whereas FOURLAWSHILL TOP NY9083 ("four hills hill hill"), a good example of onomastic tautology, is on the left. SWINE HILL NY9082, on which the Roman camp is situated may have been one of them, signifying either "boar" or "creek"; the distinction between the two has been lost in Northumbrian. There are plenty of small streams around, and there were wild boar here in early medieval times, so either would suit. The area on the right as you come up to the crossroads is called COCK PLAY NY9082 and probably designates an area where wild birds (OE *cocc* "wild bird") were wont to display. BUTELAND FELL NY8882 and the farm of BUTELAND NY8781 are further over to the west. These names are usually explained as meaning "Bōta's land", and although there was a lady of this name who owned land in the south of England, the first element is just as likely to be OE *būte(n)* "outside". But outside what? As there is a cluster of British forts around here, a tentative explanation might be that this was land outside English or Anglian jurisdiction.

Over to the east, the minor road off the A68 takes you to SWEETHOPE NY9581 (*Suethoppe* 1280 = OE *sweote* "sweet, pleasant") over the Wannies (LITTLE WANNEY CRAG NY9283, GREAT WANNEY CRAG NY9383) and the headwaters of the Wansbeck. It is tempting to link the name of the River WANSBECK (= *Wenspic* 1137, etc.) with the name of these crags, especially as the second element may be ME *pik / pike* "summit / cairn", but if this is correct we would have to assume transfer of the name from the hills to the river, and an odd vowel change, and we should still not have a satisfactory explanation for the name WANNEY. The crags appear to be an outcrop of the Whin Sill (a horizontal layer of dolerite), and *whin* (= ME *quin*) "hard stone" (*whin* "gorse" is hardly the source) may provide a clue to the name, but its etymology is also obscure. This is an intractable problem at present. The WANNEY BYRE NY9383 is a natural fissure in the rock where it is said that livestock were kept (therefore OE *bȳre* "cowhouse"), providing a possible link to the droving activity so characteristic of this area before the modern era.

From this crossroads, Dere Street with a Roman milestone on COMB HILL NY9181 (OE *camb* "long narrow hill") strikes south towards Stagshaw over the blind summits which delight drivers and car-passengers of tender years. Just below the sawmill on the left, HIGH PITHOUSE NY9180 is doubtless named after the old coal workings on FELTON HILL NY9180 (*Fyleton* 1245) behind. This hill name probably reflects a personal name such as *Fygla*, or OE *filepe* "hay", rather than OE *feld* "field". The TONE INN NY9180, formerly the *Tone Pitt Inn* according to Tomlinson, is now no longer a regular inn; its name is shared with TONE HALL, TONE COTTAGE, TONE LANE, TONE WELL (all seemingly modern and located in NY9080) and the intriguingly named TONE THRASHER NY9079, but I have no explanation for the use of *thrasher* for this derelict farmhouse beyond noting it as a variant of *thresher* "one who threshes". The name TONE appears as *Tolland* in 1182, so the original name had two elements, the second of which is almost certainly OE *land*, and the first may be either OE *toll* "tax, duty", a personal name *Tolla*, or perhaps ME *tow* "flax", giving "land on which tax is paid", "Tolla's land" or "flax land" as three possible meanings. Just north of the Tone Inn is a laneway on the left to WHITESIDE NY9180 and WHITE HOUSE NY9280 below CATS ELBOW NY9281 (another reference to wild cats) and COWSTAND HILL NY9381 (compare other similar names, e.g. BEEFSTAND HILL NT8213 which reflect the use of the land for upland grazing in the summer months). Nearby is the DRY BURN (*Drieburn* 1182), which like the nearby REED SIKE NY9380, is often dry in the summer.

The A68 crosses the CARRY BURN, which flows past CARRYCOATS HALL NY9279 (*Carricot* 1245) and into COLT CRAG RESERVOIR NY9378, just to the south. One suggestion is that CARRYCOATS is derived from Welsh *caer y coed* "fortress in the wood", but the early forms suggest rather that it is the cottage (ME *cote*) by the Carry (burn). Like the CARRY BURN NT6502 on the border with Scotland, the river CAREY in Devon, the CARY in Somerset, the Welsh rivers CAR and CERI, as well as the Gaulish river-name CARUS (now CHER / CHIERS), we are dealing with a recognized pre-English river-name. Perhaps it is related to the root found in British / Gaulish **karro*- "chariot" (whence Latin *carrus* and also English *car*, etc.) and Welsh *cerdded* "to move, walk", and therefore meant "moving, flowing". Taking a detour over to the west, the name of the burn itself occurs again in HIGH CARRY HOUSE NY8679 (LOW CARRY HOUSE is in NY8578 by the North Tyne). Nearby CAMP HILL NY9176 is one of several places with this name reflecting a hill topped by a British fort (OE *camp* was used in this way in the south of England too). There is certainly a cluster of settlements of probable late Iron-Age date in this area. Other camps have names which are less than straightforward. Of GOODWIFE HOT NY8778, the site of a fort near Birtley, we can be fairly sure that the second element is OE *holt* "wood" (despite the name having been bowdlerized recently to Goodwife Hut). Indeed, remains of a wood are still there on the site, even though there are only faint outlines of the buildings. That being the case, and given the widespread tendency to duplicate place-name elements, we could (using the magic wand of popular etymology) just about derive GOODWIFE from a first element Britt. **kaito*- "wood, forest" (Welsh *coed*, etc.) and a second element *baedd* "wild boar", as in *coedfaedd* "wild boar", so the whole name would mean "wild boar wood wood" with the *wood* element being expressed twice. On the other hand, perhaps we should try not to be too clever: PASE lists one Godwif, the daughter of a lady Leofrun who owned land in Hertfordshire, so there is no reason why the name should be unique. GARRET HOT NY8681 across the North Tyne from Redesmouth is another site the origin of which is said to be British, but the second element is OE *holt* "wood" as before and the first may simply be a personal name, although names in GARRET- do seem to be associated with Romano-British sites surprisingly often. COUNTESS PARK NY8780 is according to Watson "named after one of the many Celtic variants for a Hill" (1970:18), but which word might he have had in mind? The late Professor Jackson has shown that a British **kuno*- "high" never existed. However, there is a British place-name *Cunetio* (PNRB:328) which is probably to be identified with the Roman town at MILDENHALL SU2169 in Staffordshire, and underlies the name of the Wiltshire River KENNET. It may also explain the River KENT of Cumbria (older *Kenet*) and the CYNWYD of Merioneth, so it is just possible that our *Countess* reflects an old name of either the River North Tyne, or one of the tributary burns (the PRESTWICK BURN and the BLACK BOG BURN join together as the HEUGH BURN and flow into the North Tyne near here). Alternatively, and some may argue more plausibly, as this was part of the Percy estates, the park may have been named after a Countess of Northumberland. Perhaps the name of the PRESTWICK BURN refers to one of the Romano-British enclosures (OE *wīc* from Latin *vīcus*) in this area which was later associated with priests or monks (OE *prēost* "priest"). Like *-wick*, the *-chester* of ROUCHESTER FARM NY8977 harks back to the Romano-British period (the first element is probably OE *rūh* "rough", as in ROUGH CASTLE NS8479, a Roman fort on the Antonine Wall). The BLACK BOG BURN on the other hand seems to be modern and is named after the bog through which it flows. Northumberland names containing *Heugh* usually reflect OE *halh* in the meaning "narrow valley" (the DEVIL'S LEAP is not far away NY8679) and then "alluvial land next to a river". The *clints* in HEUGH CLINTS NY8780 probably refer to the rocky outcrops and, although the word derives from Scandinavian, it is not attested in English until ca. 1400 and so does not provide good evidence for Scandinavian settlement in the area. British origin has also been claimed for CATREEN NY8878 on the basis of Welsh *cadair* "seat", borrowed from Latin *cathedra* but appearing in place-names such as CADDER NS6172 associated with a hill-fort. However, given the number of places named after wild cats in the area, OE *catt* could equally well be the basis of the first element and EMnE *rean* "water-furrow; strip of land" could be the origin of the second.

Indeed, most of the other place-names on the eastern bank of the North Tyne are unambiguously English. BIRTLEY NY8778 (*Birtleye* 1229) was motivated by a bright clearing (OE *beorhte*), and BIRTLEY SHIELDS NY8779 by summer huts (ME *shēles* would be plural, but BOG SHIELD NY8979 and LOW SHIELD NY8880 contain the singular form of this word). BLINDBURN NY8678 is another name signifying a stream which does not flow all the time. THORNEYHIRST NY8678 means "thorny wooded hill" (OE *þornig* + OE *hyrst*); GOLD ISLAND NY8677 in the North Tyne may be a place where marigolds (OE *golde*) grew. WARKSHAUGH BANK NY8677 lies opposite the village of Wark (see below) which therefore provides the origin of the name. THREEPWOOD NY8563 means a wood of disputed ownership (OE *prēapian* "to rebuke, reprehend"). NIGHTFOLD RIDGE NY8977 has a modern ring to it: a nightfold is clearly an enclosure for stock on what seems to have been a droving route to Stagshaw, although the word has no great currency or antiquity, and the form *ridge* is modern. The name RUBBINGSTOB HILL NY8978 is probably of no great age either (perhaps the cattle used this "stump", OE *stubb*, to rub themselves), but neighbouring RUSHEY LAW NY9078 has one of the older words for hill with ME *rushy* "with rushes". The early forms of COWDEN NY9179 — also MIDDLE COWDEN and LOW COWDEN NY9178 — (*Colden* ca 1250) point to a cool (OE *cōle*) valley rather than an origin involving cows (OE *cū*). This is despite Low Cowden's location on the COWSTAND BURN NY9078, which seems straighforward, although *cowstand* is not recorded in the OED as a word. On the other hand OE *col* "coal" is not out of the question in view of the proximity of the COAL BURN, another tributary of the GUNNERTON BURN, and nearby PIT HOUSE NY9976, easily explained in view of all the old coal workings in the area, on the other side of the MALLOW BURN (OE *maluwe* "mallow" used in dyeing).

BLACKHILL FARM NY8876 is perhaps also named after local outcrops of coal. There are no old forms for DINLEY HILL NY8877, just north-west of Pit House, but it could be derived from OE *dūn* "hill" and either OE *lēah* "clearing" (as in HORSLEY) or OE *hlāw* "low hill" (as in KIRKLEY NZ1576), the latter used pleonastically. The DINLEY BURN falls into the North Tyne and flows past the intriguingly named COMOGAN FARM NY8776 for which Watson (1970:18) suggests *Caer Mogon* 'the Stronghold of the God Mogon' but there is no evidence to support this. If the name is British, the first element **kagro-* would typically usually realized as *Car-* (as in CARLISLE, etc.), so we might think of **kumbo-* "valley" (Welsh *cwm*, as in CUMWHITTON NY5052, etc.) after which the most likely element would be the name of the watercourse, for which I have nothing to offer beyond an uncertain comparison with the name of the River OKEMENT in Devon.

The village of CHIPCHASE (*Chipches*, 1229), now deserted, probably predates the castle NY8875 and the early forms are not awfully helpful in establishing an etymology. It is most likely Chip's (OE *Cippas*) hunting area (ME *chace*, earliest textual example ca. 1440). Connections with markets (as in *Cheap-side*, etc.) or with beams (OE *cipp* "leg, trunk, weaver's beam") are probably to be deprecated. CHIPCHASE MILL NY8874 is post-medieval, and onomastically transparent, but CHIPCHASE STROTHERS NY8874 are marshes (ME *strother* "marsh", from OE *strōd* "marshy land overgrown with brushwood"). GUNNERTON village NY9075 is just to the south and very much in existence. The early forms (e.g. *Gunwarton* 1169) point to the enclosure or farm of Gunner (there are two individuals with this name listed in PASE, all late 10th-century). The Gunnerton Burn falls into the North Tyne near BURNMOUTH COTTAGES NY8974, appropriately enough. BARRASFORD NY9173 (*Barwisford* ca 1250) is the ford of the grove (OE *bearwas*) and a ferry was marked on the map until the mid 1970s just downstream from the site of the old mill. The road running north-east out of Barrasford now bears the name Chishill Way, a case of popular etymology. CHISELWAYS FARM (OE *ceosol* "gravel" + OE *weg* "road") is now thought to have been built over on the eastern side of Barrasford and the 1838 Tithe Commutation Map lists CHESHILL WAYS NY9274 (*Christhill Ways Farm* in the schedule!) as having been occupied by one Thomas Thompson, along with DUNSHAW FARM NY9274 "hill copse", and the lands shown are between the crags where Barrasford Quarry has now been reopened and the road. The HERMITAGE NY9374 (*Armytage* 1496), reputed to be the haunt of St John of Beverley, is on the other side of the SWIN BURN which leads us up to SWINBURNE CASTLE NY9375, GREAT SWINBURNE NY9375 and LITTLE SWINBURNE NY9477 on the other side of the A68. These are generally taken to be compounds of OE *swīn* "pig, (probably wild boar)", which gives MnE *swine*, and OE *burna* "stream", although it is equally probable that we are dealing with a pleonastic first element OE *swin* "small stream, creek" (although this word has died out in MnE). Just to the north is REAVER CRAG NY9375, which Tomlinson (1888:211) spells *Reiver Crag* and on which he notes one of the "ancient camps" in the area. (REAVERCRAG NY9374 also marks some buildings.) A *reiver* (a common local spelling of *reaver*) is a marauder or robber and this is probably the origin of the name, although a connection with early OE *geroeba* "reeve" has been suggested (note that RIVERTON in Devon developed from *Reveton* 1238 by popular etymology). Little Swinburne Reservoir is separated from Colt Crag Reservoir by FOLLY MOSS NY9377, a bog or fen (OE *mos*) either with foals (cf. MnE *foaly* "in foal") or with a folly used in the meaning of a stand of trees. Both these senses are 19th-century. Several miles to the north-east, THOCKRINGTON NY9578 (*Thokerinton* 1223) is the enclosure of the family (this is usually the significance of OE *-ing-*) of Thoker. There is a verb OE *pocerian* "to run about", so this may be a nickname.

Above Colt Crag Reservoir is LOUSEY LAW NY9278, which may be named because the hill was infected by lice of some description (OE *lūs* "louse", ME *lowsy*, words which are used to refer to other parasitic insects as well as lice), or by plants of the genus *pedicularis*; alternatively, *Lousey* may be the popular etymology of a word referring to the small loughs which preceded the reservoir (ME *louh*, possibly ultimately a loan word from Britt. **lukku-*, which gives Welsh *llwch* "lake") with the *-y* suffix, i.e. "lakesy hill". The settlement of PLASHETTS NY9681 might be similarly motivated. The usual explanation is that the settlement was *plashed* or "enclosed by a woven fence" but this does not account for the ending or the retention of the unstressed vowel. On the other hand, a *plashet* is listed in the OED as "a little plash" or marshy pool, from Middle French *plaschiet* "small pond", which might suit the location of Plashetts near the headwaters of the Wansbeck (or anywhere in Northumberland indeed!).

The *-ing-* suffix mentioned above also appears in LITTLE BAVINGTON NY9880 (*Parva Babington* 1242), the enclosure (OE *tūn*) of Babba's family (intervocalic *-b-* appears as *-v-* also in *Averwick*, i.e. ABBERWICK NU1213). BAVINGTON HALL was a residence of the Shaftoe family, and GREAT BAVINGTON NY9880 was more significant in times gone by. DIVETHILL NY9879 is probably a hill where *divets* were got, larger versions of the bits of turf familiar to golfers (early ME *divet*, but there are various spellings) which were used for roofing. Both elements suggest this is a relatively modern name. Just to the north-east, CLAY WALLS NY9879 is redolent of other early modern construction practices. Back towards the road are COLWELL NY9575, pronounced [ˈkɔlǝl] with the loss of [w] as in Berwick etc., which denotes a cool stream or spring, the same spring line possibly being reflected in the name of nearby FAIRSPRING (FARM) NY9974, and in ROBIN HOOD'S WELL NY9574, WELL HOUSE NY9674, and HIGH WELL HOUSE NY9774. To the north of this is HALLINGTON RESERVOIR with LIDDELL HALL NY9675 on its south-western corner (the Liddells are documented as a notable Northumberland family, whose name derives ultimately from OE *hlȳde* "torrent" + *dāl* "dale, valley", as in the River LIDDEL in Cumbria), and FAWCETT NY9676 on its northern bank ("multicoloured (hill)side"). The name of FELL HOUSE NY9576 is probably transparent, but we should bear in mind that Nthb. *fell* signifies a lower, more rounded hill, perhaps just moorland. The older forms of HALLINGTON itself NY9895 (e.g. *Halidene* 1247) point to a meaning "holy valley" (OE *hālig*) and, given the number of sacred wells and springs recorded in pre-Anglo-Saxon Britain, its origin may well be linked to Colwell and the others in the vicinity. To the west of the road on a south-facing slope is BEAUMONT HOUSE NY9572 (*Beaumont* 1232, *(de) Bello Monte, Beumond, Bemound* 1296) which shares its name with houses in Cumbria, Essex and Lancashire, as well as many in France.

Driving further south along the A68 into the dip, we cross what is to modern ears the wonderfully-named ERRING BURN. However, the early forms (e.g. *Eriane* 1479) suggest a British origin in **arganto-* "silver" (Welsh *arian* "silver", and several continental rivers such as L'ARGENCE from Gaulish *argantiā*). ERRINGTON NY9571 is not actually on the burn itself, as Ekwall suggests, but is indeed the enclosure (OE *tūn*) named after it. The turn to ERRINGTON RED HOUSE NY9771 is further down the A68, but ERRINGTON HILL HEAD NY9669 is accessed from the B6318 (the so-called Military Road, of which more later). Opposite the road leading to Errington itself is a turning to BINGFIELD NY9772 (*Bingefeld* 1191), the field (whatever OE *feld* meant precisely) of Bynna's people (PASE records three people with this name). BINGFIELD COMBE NY9872 appears to be on the ridge suggested by the name, and all the buildings at this derivative settlement are 19th-century, as they are at BINGFIELD EAST SIDE NY9873, NEW BINGFIELD NY9873, and BINGFIELD EAST QUARTER NY9872. BEUKLEY NY9870, with the modern Stagshaw transmitter mast towering above it, is a difficult name as it is recorded in two types. One of these (e.g. *Boclive* ca 1250) suggests a compound of OE *bōk* "beech" (as in Old High German *buohha* — the OE *bēce*, which gives modern *beech* is a later development) and OE *clif* "rocky outcrop, cliff"): this might work except that beeches would be unusual at such an exposed location, and the modern reflex of the OE word is unusual (we should expect *Bok*- or *Buck*-). The other type (e.g. *Bokeley* 1296) suggests an original clearing, and a possible first element Britt. **boukkā* "cow" (= Welsh *buwch*, etc.), which is a fair description of the use of the land even today. Across the A68 is BEUKLEY COVERT, indeed the name may originally refer to this Iron-Age site, the Romano-British farmstead, 750m north-west of GROTTINGTON FARM NY9769, early forms of which (e.g. *Grottendun* 1160) suggest that it is to be resolved as "Grotta's hill or fort". Further off the road from Beukley we come to ROSE'S BOWER NY9971 (see below for another, very similar name), which probably reflects OE *būr* "cottage", the first element being either a proper name, or even the name of the flower (OE *rose* is a loan from Latin *rosa*), or perhaps the rushes so characteristic of the area (as in ROSEDEN NU0321 "rush valley" reflecting OE *risc, rysc*). There must also have been an early form of OE *hors* "horse" with metathesis (i.e. *hros*, compare Old High German *(h)ros*, etc.) and this would also have provided a basis for the first element as in ROSLEY NY3245 in Cumbria. One final suggestion is Britt. **rossa-*, which we find in Welsh *rhos* "moor, downland" as in ROS CASTLE NU0825 — in other words, as in so many other cases we're not short of ideas, just compelling evidence! TODRIDGE NY9971 reflects the local name for a fox (ME *tod* exists beside the modern word, as FOX COVERT NY9973 / NZ0072 demonstrates) but the second element seems to be a more modern or modernized name by comparison with the *rigg* of WHITE RIGGS NZ0073 just to the north-east.

GRINDSTONE LAW NZ0073 takes its name from the hill where millstones or grindstones (ME *grinstone*) were found or manufactured (there are remains of old workings around the summit), and then gives its name to the modern farm of GRINDSTONELAW NZ0073. Driving further south down the A68, we pass LITTLE WHITTINGTON NY9969 (*Parva Witington* 1242) on the left with the larger GREAT WHITTINGTON NZ0070 (*Great Whytington* 1296) further east and WHITTINGTON WHITE HOUSE NZ0172 on the high ground beyond. The first place to bear this name was the enclosure of the *Hwitinga*, or Hwita's people (five people with this name are recorded in PASE, alas none associated directly with this part of the world). Nearby is a farm with the curious name of CLICK'EM IN NZ0072, all the stranger as it is not unique. So far, no authority has suggested a meaning and the best I can do here is draw attention to the other places (e.g. CLICKEMIN NZ1772, CLICK-EM-INN FARM NZ1644 and CLICKIMIN BROCH near Lerwick on Orkney HU4640). Luckily, we have more information for nearby SHELLBRAES NZ0071 (recorded as *le Schellawe* 1479), so it is "the hill" (i.e. OE *hlāw* later replaced by *brae(s)* "brows") "with the summer hut". It is doubtful whether THE WHIGGS NY9970 has any political or religious significance (*Whig* is also an old term for a Nonconformist according to Heslop), but is more probably derived from Nthb. *whicks* "young hawthorn plants". HIGH BAULK NZ0070 seems to mean an elevated piece of unploughed land (OE *balca*), possibly used as a dividing line.

We have now arrived at the junction with the Military Road, begun by General Wade in 1746 as part of the campaign against the Jacobites, which follows the course of Hadrian's Wall at this point. (Readers should note that this is not the same as the *Stanegate* "stone road", the Roman road which ran over a more southerly route between Corbridge and Carlisle.) The buildings just to the south-west of the roundabout are marked PORT GATE and the farm to the west PORTGATE NY9868 (*Portyate* 1278), probably reflecting Latin *porta* "gate" and OE *geat* "gate", as there was clearly a break in the wall. The flat ground at the top of STAGSHAW BANK was the site of a fair for many centuries, held on 6 May for cattle and sheep, at Whitsun for horses and cattle, on 4 July for horses, cattle and sheep and on 26 Sepember for cattle and sheep. A further fair for sheep alone was held on 5 August. Many of the drove-roads from Scotland therefore converge on this point. STAGSHAW HIGH HOUSE NY9767 is behind what used to be an inn, while STAGSHAW HOUSE is further down the hill opposite CHANTRY FARM NY9866, which may have been the site of the medieval Stagshaw Hospital. A chantry (ME *chaunterie*) was an endowment to ensure the singing of daily mass for the souls of the departed and was applied to the clergy or the building. The name STAGSHAW (*Stagschaue* 1296) indicates a copse where deer (ME *stagge*) were to be found, but there is a variant with a first element ME *stain* "stone" (*Stainscau* ca 1340) and these were pronounced locally as [ˈstadʒɪ / ˈsteɪnʃə]. (*Stagey Bank Fair* is a local phrase for a complete mess, a rough Northumberland equivalent of *Muldoon's Picnic*.) SHAWWELL HOUSE NY9866 is named after a stream in a copse, but the building itself is relatively modern. Drivers will have noted a sign to AYDON CASTLE (*Ayden* 1242) NZ0066, clearly named after the valley. Later forms (e.g. *Hayden* 1242) suggest the first element may be OE *hēg* "hay", but, if the initial <h> is not historical, it could be OE *ēg* "island", which is not impossible as the castle is situated between two streams and is, in any case, partly encircled by the Cor Burn.

Nowadays, drivers naturally gravitate to the new A69, which bypasses CORBRIDGE NY9964 (*Corebricg* 1040) with its narrow bridge (OE *brycg*). For the first element *Cor-*, we should note that the name *Coria* appears in the Vindolanda tablets, may be connected with Britt. **koryo-* "army" (W *cordd* "tribe, troop") and has been glossed "hosting place". However, it is generally recognised that the form *Corstopitum*, often quoted in accounts of Roman Britain, is of little probative value. There are plenty of names with a first element *Corio-* (e.g. *Coriovallon*, which becomes *Cherbourg*), but the rest is a puzzle. We may ponder this as we explore the by-ways to the west of STAGSHAW BANK. Just north of the A69 is HAMPSTEAD NY9865 which reflects OE *hāmstede* (= MnE *homestead*) "hall, residence", and is therefore the onomastic cousin of more famous Hampsteads. The [p] is a later development. There is no reason to suppose that FAWCETT HILL COTTAGE NY9767, named after the multicoloured hillside to the north-west, has a different derivation to FAWCETT NY9676 mentioned above. BLACK HILL NY9767 seems transparent as a place-name, as does HOLLY HALL NY9766 if this is a modern name, but the second element of SANDHOE NY9766 (*Sandho* 1225) "sand ridge" perhaps requires explanation (OE *hōh* "ridge, height"). The name was pronounced ['sanda] locally. Just to the west is BEAUFRONT CASTLE NY9665 (*Beaufroun* 1356) with an obviously French name (*beau* "fine" + *front* "brow, facade") extolling the virtues of the house or its situation. It was pronounced ['biːvrən] in the 18th century. BEAUFRONT RED HOUSE NY9765, BEAUFRONT WOOD HEAD FARM NY9566 and BEAUFRONT HILL HEAD NY9666 are the principal locations with derivative names in the immediate area. Driving west, we come to ANICK, pronounced ['ɛɪnɪk], NY9565 (*Æilnewic* ca 1160), the farm of Æþelwine or Egelwine (an 11th-century Bishop of Durham), and ANICK GRANGE NY9565 (ME *graunge* "grain repository"). Further along the road is BANK FOOT NY9565 (i.e. "bottom of the slope"), and BRIDGE END NY9564, both names of very common occurrence. In fact, there is another BRIDGE END NY9166 just a mile or so to the west, with OLD BRIDGE END NY9265 in the middle. Clearly, this was an important point at which to cross the River Tyne.

CHAPTER THREE
3. From Hexham to Kielder

Around AD 150, the geographer Ptolemy recorded the name of the TYNE as *Tina*, probably ['tiːna]. Students of the English language will have little difficulty in tracing the development of this form via the Great Vowel Shift ([iː] to [aɪ]) and loss of the final syllable to its modern English equivalent [taɪn]. There is no sure evidence of any other names for the two branches now known as NORTH TYNE and SOUTH TYNE. Symeon of Durham (ca 1130) mentions the latter as *Tina Australis*, so this nomenclature is probably quite old too. What might the name TYNE mean? Ekwall derived it from an Indo-European root **ti-*, and a general meaning of "water, river", but this should not be taken as implying that reflexes of the root are common. Apart from the name of the River TILL (see below), the closest possible parallel is Old Church Slavonic *ti-na* "mud". The other possible comparative forms, Old Norse *þíðr* "melted" (from earlier **ti-tós*), OE *þīnan* "to become moist", and Greek *tī-los* "diarrhoea", all have watery connotations, and show suffixes with *-n-* and *-l-*. There are a number of water courses whose names may reflect this formation. Apart from the River Tyne in Scotland (the village of TYNINGHAME NT6079 is located on it), Ekwall lists TINDALE NY6159 and TINDALE TARN in Cumbria (which show shortening of the vowel). The root may also be in the name of the River TILL (*Till*, ca 1050), which flows into the Tweed at TILMOUTH NT8742 (*Tyllemuthe* ca 1050) (there is also a River Till in Lincolnshire), showing the same suffix as the Greek word. Perhaps these were named after muddy stretches of the river; the lower tidal reaches of the Tyne would be the first view of the river an incoming traveller would have. If this etymology is correct, the name of the Tyne would date back to a period not just before the English, but before the Brittonic speakers who preceded them, and have its origins in the earliest linguistic stratum, of which we have only the most tenuous evidence.

From the north bank of the Tyne near Hexham, we take the A6079 for our journey along the River North Tyne towards the head of the valley beyond Kielder. The first place we encounter is ACOMB ['jɛkəm] NY9366, but do not be seduced into thinking that this is "valley of the oaks". The early forms (e.g. *Akum* 1268) suggest that this is the dative plural of OE *āc* "oak" used in the sense of a locative, so the name means "by the oaks". The village gives its name to ACOMB FELL NY9568 to the north-east. ACOMB BUILDINGS NY9467 and ACOMB FELL FARM NY9567 are modern derivative names. And OAKWOOD NY9465 has the corresponding modern form. Just to the south is THE RIDING NY9365, which may reflect the first sense given in the OED, namely a way or path cut through or adjacent to a wood for horses and their riders; alternatively, OE *hryding* is a common term for "a clearing" and would also account for the name (nothing to do with the old *riding(s)* of Yorkshire and Lincolnshire, which are derived from ON *þriði-ungr* "third part"). SALMONSWELL NY9466 cannot be really old, as the first element is ME *saumon* (from French and ultimately Latin *salmo*) and the second may refer to a stream or a well, but, if it is the latter, *Salmon-* will probably be a personal name. NEW RIFT NY9466, the farm opposite Salmonswell, probably indicates "newly ploughed land" (early MnE *rift* "ploughed" OED). CARR HILL NY9567 and FERN HILL NY9567 are further up the road beyond the SILVER HILL NY9467 crossroads. CARR HILL may contain a personal name, or it may reflect OE *carr* "rock", a word found only in Northumbrian texts, and therefore thought by some to be borrowed from Brittonic (compare W *carreg* "rock"). Likewise, FERN HILL may contain OE *fearn* "fern", but there is also an OE *fiergen* "wooded hill" which could account for the modern name; and SILVER HILL may just contain the word for the metal, or refer to a characteristic colour (maybe the rocks or the grass), or yet be a personal name such as *Sigewulf* (as in SILVERSTONE SP6644 in Northamptonshire, where the first element became Silver- by popular etymology). Turn left at the Silver Hill crossroads for CODLAW HILL NY9468 and CODLAW DENE NY9468. The first element of these is probably not ME *codd* "bag" as one commentator suggested (what would the motivation be?), but perhaps a personal name *Cod(d)a* which has been suggested for a few place-names (e.g. CODDENHAM TM1354 in Suffolk, or CODHAM TQ5888 in Essex), or even OE *ceald* "cold" with vocalization and deletion of the [l], a process characteristic of the local dialect. Just to the west we find the intriguingly named WRITTEN CRAG NY9368, the second element of which is reasonably clear (ME *crag* "rocky outcrop"), but the first has been obscured over time. There are no early forms, but we might be tempted to posit OE *tūn* "enclosed farm" for the *-ten* part of *Written*, or, as there is no settlement here, OE *dūn* "hill". The first element, or even the entire name, could then be *ridd(en)* "cleared" (as in *Ridley* "cleared clearing", etc.). It was probably the crags near PLANETREES NY9369 that supplied the name for CRAG HOUSE NY9269. Incidentally, plane trees would be unusual for Northumberland, but the name seems clear enough, as do those of FALLOWFIELD NY9268, WESTFIELD HOUSE NY9267, and HALFWAY HOUSE NY9367 (not quite) halfway between them.

To the west of Acomb is BROOM PARK FARM NY9266, probably named after the broom growing there. LOW BARNS and HIGH BARNS NY9267 are just to the north. These names are relatively modern. Over the other side of the river is WARDEN NY9166 and HIGH WARDEN NY9167 both on the slopes of a hill from which they take their name (*Waredun* ca 1175 = OE *weard* "watch" + *dūn* "hill"). This represents a very common place-name type (compare WATCH HILL NZ0577, etc.), although the first element could just possibly be the prefix Britt. **wor-* "upon" (= Welsh *gor-*, etc.), in which case the name would mean "on the hill". THISTLE RIGGS FARM NY9167 is on the slope towards the river and the course of the STANEGATE Roman road is to the north. This represents OE *stān* "stone" + ME *gate* "road" from ON *gata*, (not to be confused with OE *geat* "gate"). The first element could well have been influenced by ON *steinn* "stone". WALWICK GRANGE NY9069 is by the river, and WALWICK NY9070 "wall farm" itself is north of this (OE *w(e)all* "wall"). In fact, WALWICK probably retains the original force of *wīc*, from Latin *vīcus*, which signified a civilian settlement outside a Roman fort, and CILURNUM NY9170 is the fort in question here, built on Hadrian's Wall to protect the Wall at the point where it crosses the North Tyne. The name of the fort appears in Modern Welsh as *celwrn* "bucket, pail" (= OW *cilurnn*) and it is therefore taken to refer to the deep natural pool reportedly now known as the Inglepool, the old name of the pool being subsequently transferred to the fort. It seems probable to me that the name CILURNUM survives in CHOLLERFORD NY9171 and CHOLLERTON NY9372 just up the river (e.g. *Choluerton* ca 1175, *Chelverton*, *Chelreton* 13th century), with second elements OE *ford* "ford" and OE *tūn* "enclosure" respectively. The problems with the precise nature of the phonological development could be avoided if we assume a degree of popular etymology, perhaps given its location in a valley, with OE *ceole* "throat, gorge" as in CHILGROVE SU8214 in Sussex.

Hadrian's Wall is such a significant structure, even today, that it is worth remembering that the word *wall* is not Germanic. OE *w(e)all* is a loan from Latin *vallum* "palisade", both with a short [a], and all the forms in other Germanic languages are similarly derived (German *Wall* for instance is not recorded until the 13th century and the word is noticeably absent from early Scandinavian). It is perhaps not generally appreciated that the Britons had a name for this edifice, namely GWAWL (*vocatur Britannico sermone Guaul*, "it is called Guaul in the British language" writes Nennius in the 9th century, but there are other early examples). The diphthong [aʊ] is derived from a long [aː] giving us Britt. **wāl*- corresponding to OIr *fál* "fence, hedge". The interesting point here is that the long [aː] suggests that the British form is cognate with Latin *vallum*, but not derived from it. So the modern name WALL NY9168, undoubtedly connected with the nearby Hadrian's Wall, probably has the name which the locals used before the English arrived. On the other side of the site of the old Roman bridge, drivers will see signs for BRUNTON TURRET NY9269, near LOW BRUNTON NY9270 and BRUNTON HOUSE NY9269, both at the bottom of BRUNTON BANK NY9369. These names are usually taken to be composed of OE *burna* "stream" (with metathesis, i.e. the vowel and [r] change places) and OE *tūn* "enclosed settlement". But, as there is no significant stream in the immediate location, OE *burna* is possibly used in its allied meaning of "spring".

Our route takes us left towards Chollerford and over the North Tyne. If you take the "one o'clock" turn-off at the roundabout, you come to HUMSHAUGH ['hʊmzhaf] NY9271, the early forms of which (e.g. *Hounshale* 1279, *Hounshalgh* 1307) indicate the alluvial land belonging to *Houn*, earlier *Hūna* (there are ten people with this name recorded in PASE). HAUGHTON CASTLE NY9272 (e.g. *Haluton* 1177) is the enclosed settlement on this site by the river and the name was then transferred to HAUGHTON MAINS NY9271 (with the shortened form of the plural of ME *demesne* "a farm attached to a mansion house" from Anglo-Norman *demeyne*, etc.), HAUGHTON PASTURE NY9072 (ME *pasturre*, another loan from Anglo-Norman), and HAUGHTON STROTHER NY8973 (ME *strother* "marsh"). WESTER HALL NY9172 might be thought to reflect OE *westra* "more westerly", but even though most names containing *hall* turn out to reflect OE *halh*, this really is the MnE form *hall*, as we know it was built in 1732, and *Wester* must be archaizing usage as in WESTERHOPE NZ1967. On the other side of the B6320 is LINCOLN HILL NY9071, which Watson suggests is a "straightforward corruption of Limekiln Hill". While linguists might prefer to call this development popular or folk etymology, his suggestion has some merit as there are several disused limekilns in the immediate vicinity, and the word *kiln* often appears as *kyll / kill* in the early modern period. However, as popular etymologies usually take place when the underlying word is unfamilar, some doubt remains, and there could be a connection with OE *hlinc* "ridge, slope, hill".

Driving north on the road towards Wark, we pass KEEPERSHIELD NY9072 on the right meaning "the warden's hut" (ME *keper* "warden") and HEATHERIDGE NY8972 on the left (this seems to be a modern name and the house was indeed built in the 1820s). The so-called KEEPER SHIELD QUARRY is on this side of the road just north-west of an area called THE SCROGGS NY8972 (ME *scroge*, "brushwood, underwood"). COLDWELL NY9073, on the river opposite RIVERHILL FARM NY9073, is one of several places in the county named after a cool spring or stream. The road turns to the west, past FAIRSHAW FARM NY8873 "pleasant copse" (OE *fæger* "fair, pleasant") and dips down towards SIMONBURN NY8873 (*Simundeburn* 1230), probably Sigemund's stream, even though the stream is nowadays marked as the CROOK BURN which rises on HAUGHTON COMMON (probably named after the settlement and castle) and falls into the North Tyne just above NUNWICK MILL NY8974. The widely-held view that the name refers to SIMONBURN CASTLE NY8673 (with *burn* derived from an oblique case of OE *beorg* "castle", i.e. *beorgan*) built by Simon, Earl of Northumberland, does not seem to be supported by the early forms. However, the castle provides motivation for CASTLE DEAN and CASTLE LANE nearby. The name of the CROOK BURN probably derives from ME *krōk* "bend". The buildings at NUNWICK NY8775 are relatively modern, but there was a medieval village hereabouts (*Nunewic* 1165), so named probably from ownership by nuns (OE *nunne*) elsewhere, as there is no trace of a nunnery in the vicinity.

We now take a short detour through the village of Simonburn itself, at first following the course of the Crook Burn. HALL BARNS NY8773 (an 18th-century farmhouse, but a bastle is noted in a survey of 1541) and UPPERTOWN NY8672 (another 18th-century building) are on the other bank. Both names are relatively transparent. On the other hand, TECKET NY8672 (*Teket* 1279) may be an old name. Watson says it is "clearly of Celtic origin", but the source is anything but clear. The most immediate comparison seems to be with LLYN TEGID, another name for BALA LAKE or LLYN Y BALA in North Wales, but the origins of this name are also obscure. Words in *-ket* and the like may be derived from Britt. **kaito-* (W *coed* "wood", as in CULCHETH SJ6595 and PENKETH SJ5687 in Lancashire), and the first element could be the Welsh adjective *teg* "fair", from **teko-* (cf. Gaulish *Uertecissa* "very beautiful"), so "beautiful wood". Alternatively Britt. **taksos* (cf. Gaulish *tacsos / tascos* "badger") would fit semantically — i.e. "badger wood" — but less well phonologically. Almost at the end of the road is GREENHAUGH NY8572, which suggests "green alluvial land", but its slightly elevated location may suggest OE *hōh* "height" instead. STOOPRIGG NY8472 is further to the west, and signifies a ridge where gate-posts (late ME *stulpe, stolpe* with characteristic vocalization of the [l]) were obtained. Returning to Simonburn village, we now make towards the CASTLE BURN into which the HOPESHIELD BURN "valley hut stream" flows near BURN HOUSE NY8673 (appropriately enough). On the other side is FENWICKFIELD NY8573. The name Fenwick is common enough locally, meaning a settlement near a marsh (OE *fen*), and also as a proper name. Just north are the remains of the HAGGLE RIGG NY8374 settlement, which dates to the Romano-British period. *Haggle* is probably a compound of *hag* "projecting mass of peat" plus *hill*, with *rigg* used pleonastically.

Returning to the main road, we make another detour, this time rather longer, from TOWNHEAD NY8774 (the name is probably modern and not uncommon) towards the hinterland of Wark Forest. First on the right is PARKSIDE NY8774, beside the park at Chipchase (PARK END and LOW PARK END NY8775 are similarly motivated). The RED BURN (OE *rēad* "red" or *hrēod* "reed") is on your left along with SLATERFIELD NY8674 and ALLGOOD FARM NY8574, both of which almost certainly contain family names. On the right, we have CONSHIELD NY8575, the hut or shieling belonging to Con (or similar), and LYNDHURST NY8675, doubtless reflecting OE *lind* "lime-tree" and OE *hyrst* "wooded hill" (LYNDHURST HILL is just to the south). GOATSTONES NY8474 is the next farm to the left, "once gatestones" says Watson, no doubt thinking of STOOPRIGG (see above) but without providing any evidence. We should note that *goat's-stones* is listed in the OED as "the name of several orchids, esp. *Orchis mascula* or *hircina*", so that might provide the motivation, or it may be named after the nearby stone circle (see below). On the right, we pass BLEAKLAW NY8475, probably meaning "black hill" as we are coming into an area of coal outcrops (indeed BLACK LAW NY8073 is just to the south-west), then NEWTONRIGG NY8375 "new town ridge" which is accessed from a lane to GOFTON NY8375 (*Goffedene* 1279), the other side of the GOFTON BURN in what turns out to be Gof's valley, which contains two "swampy clearings" further downstream, HIGH MORALEE and LOW MORALEE NY8476 (i.e. OE *mōriga* "swampy", rather than a reference to standards of behaviour. SADBURY HILL NY8276 looks like a pleonastic formation with OE *burg* and *hyll*, both "hill", but this does not explain the first element. The parallel with the name SÈVRES (from Gaulish *Satubriga*) is striking, so it could just possibly be Brittonic in which case *Sad-* would be derived from a personal name. CATLESS NY8375 is not a place devoid of cats, but rather the opposite: "clearings" (OE *lēahas*) where wild cats were to be found. To the south-west of RAVENSHEUGH CRAGS NY8375 (cf. OE *hræfen* + OE *hōh* "height") is a stone circle also known as THE GOATSTONES. Names in *-heugh* tend to be associated with standing stones, burial mounds and the like, and goats were associated with important locations in this area (see YEAVERING NT9229). Nearby, a track leads to GREAT LONBOROUGH NY8273 (perhaps "long fortified place", though the present buildings are 19th-century), overlooked by THE CARTS NY8373 (OE *ceart* "rough common"), TOWNSHIELD BANK NY8172 (there was a settlement here in the late Iron Age, but this name seems modern) and STANDINGSTONE RIGG NY8173 with genuine standing stones on a low ridge.

Further along the road across BROADPOOL COMMON, a modern name despite evidence of early settlement, is STANDARD HILL NY8275, for which I have no further information, and then LADYHILL NY8075, which also seems to have a modern name. The buildings here are 19th-century. WILLOWBOG NY7975 is just to the north-west, with MIDDLEBURN NY7975 (*Midelburn* 1268, "middle stream") — there is a burn flowing past the farm) and COALCLEUGH NY7974 "coal ravine" to the east, reminding us of the coal outcrops in the area. The burn which flows down from BELL CRAGS NY7773 has the same name and *bell* was used locally for a pit where the mine is sunk very near the surface, doubtless on account of its shape. BELLCRAG FLOW NY7773 is a watery moss or morass (one meaning of MnE *flow*), and is as far south as we are going on this trip. Just to the east is BYRESHAW HILL NY7672 (perhaps OE *byre* "shed") with HINDLEYSTEEL NY7472 (perhaps "hindberry clearing" with OE *stigol* "ridge") further west still. HOPEALONE NY7371 is an even more remote farm or valley. If *alone* is the same word as Modern English, then the name cannot be older than ME, as *alaan* is a ME formation. HOPE SIKE is a watercourse flowing westwards towards CADGERFORD NY7171 (possibly ME *cald* "cold" + *chere* "face"), and SCOTCHCOULTARD NY7170, another 19th-century farmhouse with a name probably reflecting ME *scot* "tax, tribute" and perhaps the personal name *Co(u)ltard* ("a keeper of colts", OE *colt* and OE *hierde*). SCOTCHCOULTARD WASTE is just to the north beyond HUGH'S HILL NY6971, of which the first element may be a personal name or simply a popular etymology of OE *hōh* like some other names in *Heugh*) and HUMMEL KNOWE NY7071 (ME *hommyl* "hornless, dodded", therefore "bare, without trees" like the many summits called *Dodd* + OE *cnoll* "small hill"). The river just to the west, on the other side of which lies the modern county of Cumbria, is the IRTHING (e.g. *Irthin* 1169, from earlier *Irtīn-*, a Britt. river-name). This may contain a form of *ritu-* (*irtu-* via metathesis), which we find in PENRITH NY5130 further south.

Several small streams fall into the Irthing on its eastern bank, including SMALL BURN (a very common name with *small* in the sense "narrow") and SUNDAY BURN (possibly ME *sonni* "in a sunny location"), which rise on the slopes of BLACK FELL NY7073 and flow through the embankments known, perhaps pleonastically, as BELL'S BRAES NY6871. HAG SIKE feeds the SUNDAY BURN, and LINEN SIKE joins the Irthing north of the waterfalls (perhaps OE *hlynn* "waterfall" provides the motivation for the first element) and south of LAMPERT NY6874, a farm of some antiquity accessed from the Cumbrian side. It is recorded as *Lythel lampard* 1291, *Parva Lamparde* 1372, so the first element of this early name is clearly OE *lytel* "little, small". The second may be composed of Britt. **landa-* "enclosure" (as in Welsh *llan* "church" and the camp on the Wall *Vindolanda*) and a personal name, something like *Bardos*. Finally, on the other side of SPY RIGG NY6875, GREAT WATCH

HILL NY7075 and RUSHEY RIGG NY7075 (ME *rushy* "with rushes"), the area known as THE FLOTHERS NY7076 and the THROSS BURN feed the COAL BURN which flows into the Irthing opposite the ruined farmhouse of SHANK END NY6876 (*shank* "projecting part of a hill"). The name *Flothers* is not well understood. If we can compare ON *flōð* "flood", it may mean "waters, floods"; if we can compare W *llythrod* etc. "mire, mud" (with Britt. initial [l] borrowed into English as [fl-]), it may mean "mires". The form *Thross* is no easier. If it is the same element as in the POLTROSS BURN further south, it may reflect Britt. **trās* "(a)cross". BLACK FELL, BELL'S BRAES and COAL BURN suggest coal outcrops and old mine workings. Another point of note concerns the two watch hills, the naming of which may date back to the Wardens of the Marches, and indeed the etymology of GREAT WATCH HILL (and many other names containing *watch* in the county) would be unproblematic were it not for other names such as WATCH TREES NY6860, which suggest a possible connection with Gaul. / Britt. *uocaiton* "lower wood" (e.g. VOUGEOT, Côte d'Or, and WATCHET ST0743 in Somerset).

Back to the east of BLACK FELL NY7073 we find GREAT BUCKSTER NY7172. If this name is formed in the same way as CRASTER NU2519, we are dealing with either a fort (OE *ceaster*, less likely) or a stell or fold (OE *ēstre*, slightly more likely) for the bucks (OE *bucca* also "he-goat"). BUCK BOG SIKE flows down towards LITTLE BUCKSTER and the TIPALT BURN (see the discussion of names along Hadrian's Wall). THE LINN NY7273 (LINN MOSS is just to the north-west, both reflecting Britt. **lindo-* "lake, bog" as in LINDOW MOSS "black lake bog" near Manchester) and GRINDON GREEN NY7273 (maybe "green hill green") are both now ruins. BLUE HEMMEL SIKE runs down past BLUE HEMMEL farm NY7475 (*hemmel* "cow-shed", but why blue?), THE HAINING NY7575 (*Hayning* 1304, MnE *haining* "an enclosure") and the WARKS BURN (of which more later). WHITE SIDE NY7576 "white hillside" is just to the north with almost the same name as WHITE HILL NY7776 above WHYGATE ['wagət] NY7776, which could well reflect OE *hwīt* "white" as well. Taken with WHITEHILL MOOR NY6978 and WHITEHILL NY6777, the farm on the Northumberland side of the Irthing, one could be forgiven for thinking that, like the planners of some modern housing estates, the namegivers herabouts had run out of ideas, but long-lying snow may be a common motivation. The second element of WHYGATE NY7776 may be ME *gate* in the meaning of "pasture", but the settlement itself is not that old and the name may be motivated by *gate* "road" as several drove-roads converge hereabouts. It is just possible that *-gate* reflects Britt. **kaito-* "wood" as in BATHGATE NS9766 (West Lothian). MARL HILL NY7776, MARL SIKE and MARL WELL may contain ME or early MnE *marll* "marl", a post-glacial deposit commonly used as a soil improver, or it could just be OE *mǣl* "mark, cross" with loss of [r]. COLDCOTES and COLDCOTES HILL (*Kaldecotes* 1279) NY7675 reflect OE *ceald* "cold" and *cot* "cottage". CROOKBANK NY7876, the name of a farmhouse built in 1818 and an area just to the west, probably reflects ME *crok* "bend", often applied to watercourses.

ROUND HILL NY7477 could be the name of any hill in Northumberland, so what makes this one special? One possibility is that it contains OE *rymed* "cleared" like ROUNDWAY in Wiltshire, although it has since been partially replanted. ROUND TOP NY7176 is probably to be explained in the same way. BUTT HILL NY7477, which gives its name to BUTTHILL SIKE, is just to the north (early MnE *butt* means a small hill, but ME *botte* "boundary" is perhaps just as likely an explanation). The problem with taking THE TRINKET NY7577 as signifying "a small watercourse" is twofold: this word is not recorded until the late 19th century in Antrim and Down (see OED) and our *Trinket* does not mark a watercourse. Instead, words ending in *-ket* often reflect Britt. **kaito-* "wood" (Welsh *coed*, etc.) in English place-names, and the first element could be derived from OE *trēo-hangra* "slope where trees grow" as in TRING SP9311 (Hertfordshire), so we would have another pleonastic formation. FELECIA CRAGS NY7277 seems to be a disused quarry, hence crags, and *Felecia* perhaps contains second element *-shaw* (OE *scaga* "copse") with the first a peesonal name *Feoluca* as found in FELKINGTON NT9444.

We return towards the road to Bellingham via STONEHAUGH NY7976, which sounds modern and indeed, like Byrness in Redesdale, the village was built by the Forestry Commission in the 1950s. The name may be older, however, and seems appropriate for its situation in a stony place by the Warks Burn. Nearby STONEHAUGH SHIELD is a 19th-century farm, which gives its name to STONEHAUGH SHIELDS PIKE NY7975. STANDINGSTONE CLINTS NY7976 are crags (ME *clint* "projecting crag") above the Warks Burn, some of which are reminiscent of standing stones. SHEILA CRAG NY7976 surely contains a popular etymology of ME *shēle* "hut" as its first element. BIRK HILL NY7876 is on the other side of the road and gets its name from the local *birk* "birch" (MnE has the palatalization, like *church* compared to northern *kirk*). The name ROSES BOWER NY7975 has been discussed in connection with the other farm of a similar name above; this and LOW ROSES BOWER NY8075 have a view of WINDY EDGE, the exposed ridge of moorland on the southern side of the burn. Continuing on the road towards Wark, we pass a turn to the left towards BLACKABURN NY7977 on the BLACKA BURN (BLACKABURN LOUGH NY7679 is a little way to the north-west), all names evoking the presence of coal measures near the surface. CRAIGSHIELD NY8077 was a hut built near HORNEYSTEAD CRAGS NY8077 (the form *craig* being a later development), but the present farm is more substantial than the name might imply. HORNEYSTEAD NY8177 itself is "a steading or homestead" (OE *stede*) on a tongue of land (OE *horna*). On the other side of the Warks Burn is FAWLEE SIKE, a small stream named after a multi-coloured clearing. Further down the road, THE ASH NY8177 is not an uncommon name, and may be just as old as the buildings, which are late 18th-century at their core. One possible derivation for LEADGATE NY8177 is "lead road", but as there is no lead-mining in the immediate area it could be derived from OE *hlid-geat* "swing gate" like the LEADGATE near Wark NY8177.

The Pennine Way follows the course of the road from The Ash to Leadgate and from there down to LOWSTEAD NY8178 "low (home)stead" by the Blacka Burn. Drivers follow the road to HETHERINGTON NY8278 (*Hetherrinton* 1291), the farm of *Hæðhere* or his sons, on a ridge overlooking the KIRK BURN, which falls into the North Tyne just upstream of Wark. The name is clear, but which kirk or church is meant? Although there are no remains to be seen now, there was indeed a chapel mentioned in documents of 1360, situated near the appropriately named KIRKFIELD NY8578. The road heads south from Hetherington towards a junction; we turn right for BURMOOR NY8277 and MORTLEY NY8277, left for THE BARNES NY8477 (this is surely MnE *barns* "covered buildings for storage") and WARK. Depending on when the name was coined, WOODLEY SHIELD NY8476 may either be a hut belonging to or near Woodley, or we could take the first name literally as "clearing in a wood". As for BURMOOR NY8277, place-names in *Bur-* usually derive from OE *burg* "fortified place" (perhaps the bastle at adjacent Mortley provides the motivation); OE *mōr* meant "upland waste" and this is likely to be the second element. MORTLEY itself indicated a clearing, but the first element is less than clear (could it be OE *mōt* "meeting" with non-historic *-r-*?).

Intrepid drivers will have reached the B6320 again by now and be prepared to continue the trip northwards. GLEN RIDLEY NY8676 (probably a modern name) is a wooded valley where the Gofton Burn falls into the North Tyne, with LATTERFORD NY8676 and LATTERFORD DOORS NY8675 just to the west. There is probably a deserted medieval village here, but no remains seem visible now; the place is probably named after the ford, but what is the first element? If we may compare LATTIFORD ST6926 in Somerset, recorded as *Lodreford* in the Domesday Book, the first element may be OE *loddere* "beggar, pauper". The problem with deriving this from OE *loddere* directly is that OE *o* does not normally correspond to Nthb. *a* (except before nasals, therefore not here). Also, the etymology of *loddere* in OED is debatable because it ignores Lat. *latro*, *latronis* "robber", which gives Welsh *lleidr*, pl. *lladron*. If we assume a loan direct from Britt. **latrō*, **latrū*, the problem with the vocalism disappears. A ford is after all a good place for robbers to lie in wait. And if that is true, we might think of *Doors* as being derived from Britt. **dubro-* "water"; alternatively it may just be OE *dor* "door" which is sometimes found in the plural in the sense of a "pass", i.e. "robber ford pass".

The next watercourse upstream is the WARKSBURN (*Werkesburn* 1293), the name providing an interesting parallel to that of Redesdale with the genitive first element in *-s*. WARK ON TYNE NY8677, often so named to distinguish it from WARK ON TWEED NT8238, is recorded as *Werke* 1279 and *Werk in Tyndale* 1294, reflecting OE *weorc* "fort, (earth)work". There are plenty of Iron-Age and Romano-British settlements hereabouts to provide adequate motivation for the name, and many have been lost such as that recorded near PASTURE HOUSE NY8577 (this name seems to be modern) on the minor road to the west out of the village. LONGSTROTHER NY8377 is no doubt named after the marshy area (ME *strother*) around the DEAN BURN "valley stream".

Nearby CROSSRIDGE NY8377 is definitely on a ridge, but was there a cross there as well? MANOR FARM NY8378 has all the hallmarks of a modern name, and was created for the Duke of Northumberland at the time of the enclosures. BROWNSLEAZES NY8379, with its so-called double plural (OE *lēahas* "clearings", which gives us names in *-leas / -leys / -leaze* etc. was already plural) suggests a later name too, so *Browns-* is probably a personal name. We cross the Warksburn near BRIDGE HOUSE NY8279 (this name derives from the bridge), passing a medieval cross on the right and the turn to WATERGATE FARM NY8179 (perhaps meaning "watery pasture", as ME *gate* could also mean "pasture") on the left just south of the Pundershaw Burn. PUNDERSHAW NY7880 is an area to the north-west of this and was probably originally a copse with a pound for animals (OE *pund*) or where the pounders (i.e. those whose job it was to keep the pound, later used as a surname) lived. HAWK HIRST NY8079 is the wooded hill where hawks were to be found (OE *hafoc*, ME *hawke*). ESP HILL NY7979 probably contains OE *æspe* "aspen".

Unsurprisingly, Wark appears in several local names: WARK COMMON is the area of originally common land to the north-west of the village; WARKSWOOD NY8478 and WARKSFIELD HEAD NY8478 seem both to be 19th-century farms. If HOUXTY NY8578 (*Houstyes* 1304) can be derived from a word meaning "hog-sty" (OE *hogges* + OE *stig*), it is likely that the name was then transferred to the HOUXTY BURN which falls into the Tyne just to the north, as it has been given to HOUXTY BANK, the bendy incline in the road. BANKHEAD NY8479 is the farm at the top. Before we reach the 18th-century farmhouse of WOODPARK NY8479, a gated track turns off to the west along the Houxsty Burn. BILLERLEY NY8479 is the first farm on the left, the site of another medieval village. This was doubtless named after a clearing, perhaps where water cress could be found (OE *billere* might be a loan from Latin *berula* "water cress", itself a loan from Gaulish, so there could have been a similar Brittonic word as well). This is certainly an ideal place for cress to grow. The road bifurcates a little further on, one fork going to SHITLINGTON HALL NY8280, the approximate site of the medieval village of SHITLINGTON (*Sutlingtun* ca 1240, *S(c)hutelington* 1279), usually taken as the enclosed farm of Scytel and his people. Alternatively, it may be related to the name of the SOOT BURN which rises on Shitlington Common and falls into the Houxty Burn between ESP MILL NY8279 (reflecting OE *æspe* "aspen") and the hall itself. SHITLINGTON CRAG NY8280 is on the other side of the SLADE SIKE (probably containing OE *slæd* "dell, valley") with EAST HIGHRIDGE NY8280 and WEST HIGHRIDGE NY8180 on its upper reaches. These farms, SHIELDFIELD NY8380 "hut field" and BLINKBONNY COTTAGE NY8380 (named after another place?) are overlooked by EALINGHAM RIGG NY8281 to the north. EALINGHAM NY8480 (*Evelingham, Evelingjam* 1279), usually taken to be "the homestead of the sons of Eofel", is back towards the road and the medieval spelling shows how early the local pronunciation [ˈiːlmdʒəm] is. NORTH BARNEYSTEAD NY8180, STONE HOUSE NY8180 and HIND RIGG NY8180 are reached from the road which leads to WATSON'S WALLS NY7881, *walls* often being a designation for a delapidated structure, but if the name was ever appropriate here it now no longer is, as the building is inhabited and in good repair. The SWINESHAW BURN, which rises near here, may get its name from a copse named after pigs, maybe wild boar, or (pleonastically) a small stream. WATCH CRAGS NY7882 and DUN'S PIKE NY7781 (probably OE *dūn* "hill") overlook the valley of the Chirdon Burn, of which more later. In the case of WATCH CRAGS, the same considerations apply as to GREAT WATCH HILL NY7075, discussed above.

On the other side of the B6320 is ANTON HILL NY8580 and one thinks immediately of a hill belonging to Anton or Ant(h)ony, but it might equally well mean "solitary farm hill" (OE *āne* "one" + *tūn* "farm"). LEE HALL NY8679 (*La leye* 1415) and LEE HALL FARM NY8680 are by the river but are probably named after a clearing. From the straight stretch of road, called the *Mile Bank* locally, which runs down towards Bellingham you get a good view of the village nestling in the valley. BRIDGEFORD NY8582, where another possible medieval settlement may have been, is by the river. There is neither a ford nor a bridge, but BOAT FARM NY8482 is on the other bank where a ferry boat used to be moored and used regularly before the bridge was built at Bellingham (the farm is reached by a road known locally as THE BOAT ROAD). It was called NORTH BRIDGFORD on the 1866 OS map. EALS NY8482 is where the EALS BURN falls into the North Tyne. An *eale* is a local word for an island, or at least slightly elevated ground, in a flood plain.

At the bottom of the MILE BANK there is a turn to the left up the south-western bank of the North Tyne, which provides the opportunity for another detour. So for the moment we do not cross the new bridge (built in 1834) into Bellingham. Leaving the delights of the onomastics of the so-called capital of the North Tyne until we have explored its hinterland more fully (you will see why as you notice all the place-names containing the element *bell*), we turn left up the southern bank.

On the southern side, we pass DUNTERLEY NY8283, named after another clearing, but there are no old forms to help us with the first element. If we can compare with DUNTERTON SX3779 in Devon, which appears as *Dondritone* in the Domesday Book (also *Duntertone* 1242), we might take it back to a British **dūno-trebā* "fort-village", but unlike its namesake in Devon, there are no obvious remains of a fortification. The cottages at WOODHEAD NY8183 are unsurprisingly at the top of the wood towards Shitlington Common. Just to the west is BRIEREDGE NY8083, presumably "briar hill or crest" (OE *brēr*), although there are no old forms. To the south of this farm are an old tilery and some bell pits, of which we will note a great many on both sides of the North Tyne.

HESLEYSIDE NY8183 (*Heselyside* 1279), set in its park with cattle grazing, provides a pleasant façade from the road, and we see that mature trees have replaced the hazels on the hillside which together provide the motivation for its name (OE *hæsel* + OE *-ig*, — the adjective *hazelly* is not recorded until the 18th century). MANTLE HILL NY8184 has no surviving medieval forms, but names ending in *-le* usually derive from OE *hyll*, so we are probably dealing with a pleonastic formation. Taking this development one stage further, the first element could be Britt. **moniyo-* "hill" (seen in MINTON SO4390, LONG MYND SO4092, etc., Welsh *mynydd*) in which case it would be a threefold pleonasm "hill hill hill" (like PENDLE HILL), but it could also be OE *(ge)mæne* "common", so "hill open to all". HESLEYSIDE MILL NY8084 lies on the haugh land by the river, and OLD MAN'S SHEEL NY8083 (another spelling of ME *shēle* "hut") on the hill above. LOW CARRITETH NY7983 (which gives its name to LOW CARRITETH BURN) and HIGH CARRITETH NY7983 (*le Caryte* 1325, *Caryteth* 1597) contain an oblique case of Latin *caritas* "charity" and the farms could well mark the approximate location of a hospital in medieval times. CARRITETH MOOR is to the south and there are remains at CROWDIE LAW NY7983 (unlikely to be the local dish *crowdy*, unless it refers to the shape of an upturned basin, or of the crowdy-cake made from oatmeal, so a connection with *craa* "crow; outcrop of coal" is a plausible alternative). CARRITETH DENE is in the valley below.

Further along the road to the north-west is SNABDAUGH ['snapdʊf] NY7884 (*Snabothalgh* 1325), the high ground to the south being SNABDAUGH MOOR. The early form shows that it is named after the alluvial ground next to the river, and there is also an early MnE *snab* "steep place of ascent, rugged rise" (with *-ot* as diminutive suffix?), which would certainly fit the terrain. BIRKS NY7784 (the local form of MnE *birches*) is on the other side of the CHIRDON BURN, which rises on the slopes of BLACK KNOWE NY6780 "black knoll" because of peat or coal). CHIRDON itself NY7683 (*Chirden* 1255) is probably named after the winding valley (OE *cierr* "bend") although the first element could conceivably be a proper name such as *Ceorra*, or OE *cir(i)ce* "church". It is reached via CADGER FORD NY7683, i.e. "cold cheer ford", but with ME *chere* in the sense of "face, slope". These days there is a bridge, which also provides vehicular access to WHITCHESTER NY7783, a relatively isolated farm on WHITCHESTER MOOR without any obvious motivation for the name (OE *hwīt* "white" is understandable, but OE *ceaster* must have been generalized as a farm name in the same way as *hop* and *tūn*). From here, you can walk back to Snabdaugh with a view of DALLY CASTLE NY7784 (1279 *Daley*), which may contain either OE *dāl* "part" (= German *Teil*), used in the compound *dāl-land* "common land", or OE *dāl* "valley" (= German *Tal*), the latter probably the case in DALHAM TL7261 (Suffolk) and on balance more likely here. Alternatively, you can walk towards DUN'S PIKE NY7782 (probably reflecting OE *dūn* "hill" + ME *pik* "summit", but DUN'S MOSS is just to the south so this may be a personal name, as in nearby CHISHOLM'S MOSS NY7781), with a view of GALLOW LAW NY7582 (OE *gealga* "gallows, gibbet") and CAIRNGLASTENHOPE NY7580. Whatever OE *hop* may have signified — "small enclosed valley, enclosure, farm" — this name remains obscure, except to point out that it refers to a small clearing in the forest and an area south of this, and that it may contain Britt. **karnom* "cairn" and **glastom* "green sward, woad" (Latin *glastum* "woad", a major source of blue dye in the medieval period, may be a loan from Gaulish or Brittonic).

Back in the valley, you can leave Cadger Ford and strike out for CHIRDONHEAD NY7181 (not the head of the Chirdon Burn, as explained above) passing BOWER NY7583 (OE *būr* just meant "dwelling, habitation" without any of the romantic connotations of later usage) and ALLERYBANK NY7481 (OE *alr* "alder" with the adjectival ending *-y*, so "slope covered in alders"). Just upstream is JERRY'S LINN NY7481, a rather pretty waterfall below JERRY'S KNOWE to the north, both names with an obscure first element. ROUGHSIDE NY7483 (*Rughside* 1357) is now used occasionally as a bothy for walkers and is named after the uneven hillside (OE *rūh*), these days covered in forest. Just to the north, EALS CLOUGH NY7483, which appears to be both the name of the stream and the steep-sided valley through which it runs (ME *clogh* "ravine"), can hardly reflect MnE *eale* "island in a flood plain", but Heslop lists *eel-beds* "the water-crowfoot" and *eel-ware* "ranunculus fluitans", both plants which could easily grow here. For intrepid explorers making for the border with Cumbria, we encounter BOLTS LAW NY6981, now a hill with communications equipment but probably always an ideal place for a dwelling (OE *bolt*). CLOCK'S CLEUGH is a ravine flowing down to the south, and although the first element is obscure it is tempting to compare with Welsh *clog* "steep rocky slope" (Britt. **kloka*, cf. Cornish *clog* and Irish *cloch*). The rocky outcrops (ME *crag*) of CHRISTY'S CRAGS NY6882 are just to the north, and a number of English place-names have similar elements indicating a Christian place, or a place with a cross (e.g. CHRISTLETON SJ4465, CHRISTOW SX8384), but there are no early forms to aid us here (and, as far as I am aware, no actual remains of crosses either).

The name of RUSHY KNOWE NY6781 with its memorial to Lord Robinson of Kielder and Adelaide (who became chairman of the Forestry Commission in 1932) is much more transparent (and common too): "small hill with rushes". WHICKHOPE NICK NY6681 probably gets its name from the pass between the hills (MnE *nick*). (We will return to WHICKHOPE itself NY6786 later when we explore Kielder Water and what lies beneath.) BLACK KNOWE ("black hill" because of the peat or coal outcrops) and HUMBLE HILL ("bare or hornless / dodded hill" reflecting OE *hamola*) NY6481 are over to the west. The area to the south of this is marked THE KNARES NY6480 (ME *knar* "rugged rock"), and there are two sets of LONG CRAGS: NY6380 and NY6582.

Returning to the Kielder road, we travel north and join the "main" road from Bellingham opposite BENT HOUSE NY7785, so named, not after the manner of its construction, but after the rough grass which is common hereabouts (OE *beonet*) (we have noted this being reinterpreted as a personal name in BENNETSFIELD NY8595). The road then swings round to the west past HOTT NY7785, which just means "wood", and GREYSTEAD NY7785 (the church is now closed). Perhaps the latter contains OE *græw* etc. "grey", but the *Grey-* element usually turns out to be something else in place-names, so OE *grāf* "grove" is more likely, together with OE *stede* "place, site of a house". Further upstream, THE EALS NY7685 really do signify the islands on the flood-plain like EALS near Bellingham. RIDLEY STOKOE NY7485 is situated below STOKOE HIGH CRAGS NY7584, but we must not be misled into thinking that the *-oe* is a relic of OE *hōh* "height", as the early forms (e.g. *Stokhalche* 1279) show that the second element is OE *halh* "land by the river". The first element is OE *stocc* "post, tree-trunk". RIDLEY is also a personal name, but derives ultimately from a word meaning "cleared clearing". The name may well have been used here to distinguish it from the STOKOE NY7386 on the other side of the river. (It is hard to say which settlement the medieval forms refer to, but they are marked as being in Greystead parish.) STOKOE BURN joins THATCHY SIKE (*thatchy* is relatively modern in standard English, but this clearly means "small stream abounding in roofing material") and falls into the North Tyne downstream.

The SMALES BURN is the next tributary upstream (hence SMALESMOUTH NY7385, both a farm and an area of what is now forest to the south-west), taking its name from SMALES NY7184 (*Smale* 1279), a remote farm just above the waterfall called SMALES LEAP (OE *hlȳp* "a leaping place" is frequent in place-names). There is an OE adjective *smæl* "narrow", and indeed the Smales Burn is narrow when compared to its neighbours, the WHICKHOPE BURN and the CHIRDON BURN. To the west of the farm, we have various ridges: BLUESTEEL RIGG NY7185 (above the BLUESTEEL SIKE, both reflecting OE *stigol* "ridge"), DROVE RIGG NY7085 (perhaps associated with a drove road), DINGS RIGG NY7084, ARCHY'S RIGG NY7083 (I have no sensible suggestions for why these ridges were so called) and HARE RIGG NY7084 (probably reflecting OE *hara* "hare") above the BROOMLEY BURN "broom clearing stream". Further west still are KYLOE CRAGS NY6983, probably "cow clearing crags" and BYRESHIELD HILL NY6783 "cattle-shed hut hill" with BYRESHIELD GRAINS NY6883 (ME *greyne* "fork, division", the word being used locally for the river branches of the North Tyne).

RIDGE END NY7285 has a self-explanatory, modern-looking name, but the farmhouse goes back to the 16th century at least (the nearby car-park is called RIGG END VIEWPOINT, so this probably contains an earlier form of the name). Like STANNERSBURN NY7286 (early MnE *stanners* "the small stones on the margin of a river"), it is accessed via the old course of the road, and lies on the lower slopes of WHITE SIDE NY7185 "white hillside". Over the other side of the road is YARROW NY7187, which probably reflects OE *gearwe* "yarrow" and maybe another element, now lost, or preserved in YARROWMOOR NY7086. The area marked as YARROW MOOR NY7086 on the older maps is now under Kielder Water. The same is true of SHILBURN HAUGH NY6986 (*Shovel-*, *Scholeburn* 1329) and EMMETHAUGH [ˈɛɪmɪthaˑf] NY6987 (*Emmoteshala* 1169) which used to lie on the haugh land where the Whickhope Burn joined the North Tyne (hence probably OE *ēa-mōt* "waters' meet", although OE *ǣmette* "ant", either as the creature itself or a nickname for the occupier, cannot be ruled out and popular etymology probably influenced the modern appearance of the name in any case). Most of WHICKHOPE NY6886 is also now under the reservoir, covering the enclosed valley or land and either the couch grass (ME *quik*) or the hawthorn (late ME *quyksette*) which grew there. BULLCRAG EDGE NY6786 has now become BULLCRAG PENINSULA (the first element may be ME *bole* "bull" or ME *bole* "tree-trunk") and it used to overlook OTTERSTONE LEE NY6787, which is obviously a clearing with what may well be stones where otters were to be seen. But by comparing names such as ADDERSTONE NU1330, this name could equally well have denoted a farm belonging to someone called Ottar, Otwin or the like (i.e. Ottar's *tūn* with the orginal [s] coming to be identified with the second element).

The new road passes just to the east of LOW CRANECLEUGH NY6685 "crane ravine" (OE *cran*), now a visitor centre, then much more isolated just below the confluence of the WHICKHOPE BURN with the HUMBLE BURN (see HUMBLE HILL above), the BINKY BURN NY6684 (perhaps *bink* "bench" + adjective ending *-y* from the large flat stones) and the BLACKCLEUGH BURN, which flows north from REEKER PIKE NY6682. The only connection I can suggest for the latter is with *reek* in the sense of "smoke" rather than "smell" and that the smoke was associated with acitivity such as charcoal burning (there is also a disused quarry there). The upper reaches of the Whickhope Burn are overlooked by GREY MARE'S CRAGS NY6182, a surprisingly common name often associated with boundary stones which brings OE *mearc* "boundary" to mind (the *grey mare* would then have arisen by popular etymology). Just to the north is GILL PIKE NY6183, also GILL HASSOCK NY6184 (OE *hassuc* "tuft of grass" seems not improbable, but I have no suggestion for the first element, unless it be ON *gil* "ravine" rendered possible by its proximity to Scandinavian settlements to the west). The burn runs past BROOMYLINN NY6384 (*broomy* "covered by or abounding in broom" is early MnE) and is overlooked by BROOMY HILL NY6385 and STOWER HILL NY6485 (perhaps local *stower* "hedge stake or post"), beyond which the LITTLE CRANECLEUGH BURN falls into the reservoir. The NISHAW BURN (perhaps "new copse stream") runs between these last two hills, rising on the northern slopes of BROOMY HILL itself. Interestingly, there are two locations called BURNT TOM. The first is an area of forest and moorland NY6286 and the second is BURNT TOM CRAGS NY5981 on the border with Cumbria. It is worth noting that *tom* is known from the east of Scotland meaning a "knoll or rounded eminence", but that this meaning may be due to a Brittonic substratum (rather than the *tom* usually meaning "copse" prevalent in the west of Scotland). In that case, we are probably talking about Welsh *tom* "mound, pile (nowadays usually of excrement)", which gives us *tomen* "hillock" as well as *twmpath* "knoll". The fact that the modern Welsh word now usually designates what is in the pile should not distract us from the probable original meaning. For *burnt*, we may compare Welsh *bryn* "hillock, knoll" from an earlier **brunnyo-* (= Irish *bruinne* "breast", another rounded eminence) and taking them together produces the now familiar pleonastic formation. It just needs metathesis (where the [r] changes places with the vowel) and analogical resegmentation (where the [t] of the second word becomes attached to the first) to produce the modern form.

The main road strikes north with KATESHAW HILL NY6586 (the first element possibly reflecting OE *cot, cyte* "cottage" or *catt* "cat") to the west past LEAPLISH (now LEAPLISH WATERSIDE PARK) NY6587. This name is thought to contain OE *hlīp* "leaping place" and ME *leche* "a stream flowing through boggy land") and the old farm at LEAPLISH overlooked MOUNCES NY6588, now inundated, and for which I have no satisfactory etymology (it was a nine-bedroomed shooting-box for Sir John Swinburne and the centre of a 17,000-acre estate). The name survives in the MOUNCES BURN which flows into Kielder Water, and in the picture of MOUNCES KNOWE by William Mulready in the Fitzwilliam Museum in Cambridge. HAWKHIRST COTTAGE NY6688 was some way to the south of the present HAWKHIRST NY6689 by the lake, at the point in fact where the HAWKHIRST BURN now feeds the reservoir (OE *heofoc* "hawk" + *hyrst* "wooded hill") below the slopes of RUSHY KNOWE NY6588 (ME *rushy*). WELLHAUGH NY6689 (probably "haugh-land with a stream / spring") was also a farm now beneath the water, but the name survives in an area of forest NY6489. The road crosses the LEWIS BURN near MATTHEW'S LINN FISHING CENTRE NY6490 where there may have been a waterfall (OE *hlynn*), but the motivation for the first elements is no longer apparent (compare DAVIDSON'S LINN NT8815, etc.)

The LEWIS BURN ['luːzbɔːrn] (*Lusbur'* 1318, *Lusburn* 1327) may contain OE *hlōse* "a pigsty", by comparison with the name of the LOOSE in Kent, which is a back-formation from the name of the village of LOOSE TQ7652, but it is difficult to justify that in terms of motivation here, as the village of LEWISBURN NY6590, now under water, appears to take its name from the burn. However, if the earthworks associated with the now inundated village are of Roman date, we might think of a Brittonic form, and one which springs to mind is the precursor of Welsh *llysiau, llysau* "herbs" from an earlier **luss-* (Middle Cornish *losow* etc. and the place-names LUSS NS3592, LUCE NY1871 and WATER OF LUCE im Dumfries and Galloway); even Britt. **lisso-* "fort, court" (e.g. as in LLYSFAEN SH8877) could be the source if we could assume confusion of [iː], [yː] and [uː] in the early period. Perhaps we should not aim so high: OE *lūs* is our modern *louse* and the word was applied to a variety of parasitic insects. It is all speculation.

The LEWIS BURN itself rises on the slopes of BLACK HILL NY5784 on the border with Cumbria and is fed by a number of tributaries. The LISHAW BURN (LISHAW RIGG NY6086 is just to the north) may reflect a reduced form of Lewis or a Norman French article (*Ricardus de la Scawe* is recorded in the Northumberland Assize rolls for 1256). The burn has its source to the north of STOTT CRAGS NY5983, and such eminences are often named after livestock, here ME *stot* "horse or (in the north) young castrated ox". LONG RIGG NY5885, GEORDY'S PIKE NY5985 and GREENS' GEARS NY5985 separate its headwaters from those of the Lewis Burn. A *gair* (or *gear*, as here) is "a strip of verdure on the uplands [...] a bright green, grassy spot surrounded by bent or heather" (Heslop), so it is just possible that the colour is reflected in the place-name rather than the surname of the owner, but the existence of nearby ROBINSON'S GEARS NY5783, or indeed STORY'S GAIRS

NY6282, makes this less likely. The YETT BURN, which seems to reflect OE *geat* "gate" (perhaps the motivation was a gate on a drove road), has its sources on BURNT TOM CRAGS NY5981 and joins the Lishaw Burn below MUCKLE KNOWE NY6185 (*muckle* "large" is the northern form of OE *micil*, OE *miccel*, which we still see in the surname Muckle, otherwise Mitchell). They join the Lewis Burn below HIGH LONG HOUSE NY6086 (LOW LONG HOUSE NY6287 is now a ruin).

The Lewis Burn is fed from the north by a series of small streams, flowing down from GLENDHU HILL NY5686 (the Gaelic appearance of this name has to be secondary, GLEN DHU is an area just to the south in Cumbria), MARVEN'S PIKE NY5787 and ELLIOT'S PIKE NY5987 (these seem to be personal names), all part of the area called CAPLESTONE FELL. There are various cairns on the top, which might be sufficient motivation for the *stone* element of this name, but the first is hardly ME *capel* "chapel" as in names such as KING'S CAPLE (Herefordshire) and is much more likely to be a compound of OE *copp* "top, summit" and OE *hyll* "hill" as in COPPULL (Lancashire). From west to east, these streams include: FOXHOLE SIKE, DROWNED SIKE, STONY SIKE, QUEEN'S SIKE (goes with QUEEN'S KNOWE NY6086), GREENSHAW SIKE, FORKING SIKE, ILL SIKE, PEAT SIKE, and GEORDY'S SIKE (goes with GEORDY'S KNOWE NY6287). They all seem to be fairly modern and in some cases transparent, although some lateral thinking is required occasionally (e.g. an *ill* is used locally for a disease of stock, so perhaps there was something in the water, or it was where stock were put to recover from minor ailments like the poorly corner of a field).

The AKENSHAW BURN, named after a "copse of oaks", but modernized as *Oakenshaw* on Armstrong's map of 1769, joins the Lewis Burn opposite a small group of cottages called, appropriately enough, THE FORKS NY6388. There are now few traces of the old LEWISBURN COLLIERY NY6388. The house AKENSHAWBURN NY6089 is probably early 19th-century (the bridge was built in 1821 for Sir John Swinburn). Like WILLOWBOG NY7975 near Wark, nearby WILLOWBOG NY5989 (*Willow Bogg* 1769) has a relatively modern name. The confluence of the NEATE BURN (*neat* "bovine animal, calf", as in *neat(s)foot oil*, connected with an old verb meaning "to use") and the GRAINS BURN (ME *greyne* "fork, division") is just to the west. These are fed by small streams such as the COAL GRAIN, the BIRKY GRAIN, and the WATCH GRAIN from the marshy area called HOBB'S FLOW NY5690 (like most apparently personal names, this probably reflects a development via popular etymology from a common noun such as *hob / hub* "thick sod, pared off the surface of a peat bog"). This is also the area of "disputed grounds" around BLOODY BUSH NY5790. The Bloody Bush Road was one of the droving routes, hence the old toll pillar on the border. Back towards the east, a track strikes north to the west of BUCK BURN under BUCK FELL NY5990 (OE *buc / bucca* "male goat / deer") and STUCKIN KNOWE NY5991 (Heslop lists *stuckin* "small pole", which is probably the same as the OED's *stucken*) with BUCKSIDE KNOWE NY5992 "goat hillside hill" on the border itself. The LITTLE BURN rises on the eastern slopes of STUCKIN KNOWE and joins the AKENSHAW BURN below WHITE FELL NY6190 (the adjective *white* is applied to so many hills in

Northumberland). FERNY KNOWE NY6289 and DINMONT LAIRS NY6290 separate the Lewis Burn from the stretch of Kielder Water now called BAKETHIN RESERVOIR NY6391. A *dinmont* is glossed by Heslop as "a ten month" meaning a male sheep after the first shearing of the new year, but the OED is less sure regarding the source of the first element, whereas a *lair* is a word for any "soft and yielding surface".

The old BAKETHIN farm NY6491 was actually on the other side of the North Tyne and is now lost along with its etymology (there does not seem to be any evidence for Watson's suggestion of "thorny back or ridge"). BEWSHAUGH COTTAGES NY6391 are also submerged, but the name lives on, somewhat inappropriately, in the area of forest to the north-west as it probably means "beavers' haugh-land" (OE *beofor*) and is reminiscent of a time before the beaver became extinct in Britain. CAT CAIRN NY6192 and CRAG HEAD NY6194 are much more in keeping, but seem more modern too.

KIELDER NY6293 (e.g. *Keilder* 1326) probably takes its name from the KIELDER BURN, which is usually grouped with other streams and rivers of a similar name, such as the CALDER in West Yorkshire (e.g. *Kelder* 1170), the Calder in Lancashire which falls into the River Wyre (e.g. *Kelder* ca 1200), and the CALDER in Lancashire which falls into the Ribble (e.g. *Caldre* 1193). There are more in Cumbria, at least five in Scotland and the Welsh stream names CALETTWR and CLETTWR, which support an etymology **kaleto-dubro-* "hard water" (Welsh *caled* "hard", but probably not in the meaning "lime-bearing" which is relatively late in Welsh, as in CLEDWEN SH8964 in Denbighshire). The second element gives W *dwfr* "water" seen in place-names such as DOVER in Kent).

For those making their escape to Scotland, the road runs through Kielder village on the western side of Deadwater Burn. BELLSBURNFOOT NY6194 is located at the bottom of BELLS BURN (which rises on the northern slopes of BLACK KNOWE NY5891) with BELLS MOOR NY6095 just to the north. The most sensible explanation for the colour (BLACK FELL NY6093 is also to the east of the burn) is the existence of coal outcrops near the surface, which were mined working "away in every direction round the bottom of the shaft like a bell", hence *bell-pit* (see Heslop). As we shall see, the *bell* element is very common in North Tynedale. KERSEYCLEUGH NY6195 is a ravine with an island where cress grew, if we are right in comparing KERSEY in Suffolk (OE *cærse* "cress" + *ey* "island"). It could simply be MnE *cressy* (with metathesis, where the [r] and the vowel change places), as adjectives in *-y* such as *ferny, allery, rushy, sighty* are very common in Northumberland place-names. KITTYHIRST NY6096 also has a problematic first element (the second is OE *hyrst* "wood, wooded hill"), but one which is surprisingly widespread, appearing as either *Kitty-* or *Kiddy-* (see the index). There is a Northumberland word *kitty* meaning "small", which would fit most contexts; alternatively, OE *cȳta* "kite; bittern" (this explanation would suit KITTY CRAGS NY9718) or OE *cyte* "cottage, cell" are possible sources. It is just possible that these names are older and reflect Britt. **kaito-* "wood" (Welsh *coed*), which was taken over by Saxon speakers to produce names in *Chet-* and *Chit-* (such as CHETWODE in Buckinghamshire, or CHITHURST in Sussex — an intriguing parallel!) and possibly **tigo-* "house" (Welsh *ty*). Of course, this explanation only works if we assume that the palatalization ([k] to [tʃ]) which took place in the south did not occur here, and sometimes it didn't (so we have northern *kirk* against southern *church*). However, there are two disused quarries on nearby Deadwater Fell which bear the name WIRCHET and LITTLE WIRCHET, both NY6297, possibly also reflecting **kaito-* (like WATCHET in Somerset, see Watch Trees above), this time with a prefix *wer- (also **war-*, Welsh *ar / gor / gwar*) "on, near". Perhaps popular etymology worked to constrain the sound laws.

DEADWATER is also the name of a farm NY6096 on the western bank of the DEADWATER BURN, so called perhaps because it allegedly runs sluggishly, but whatever the reason the etymology seems straightforward. The burn rises between DEADWATER FELL NY6297, DEADWATER MOOR NY6398 and PEEL FELL NY6299 on the border with Scotland. Perhaps there was a fence (ME *pēl*) which gave the fell its name. Following the border north-eastwards brings us to the massive slab of sandstone called the KIELDER STONE (or STANE as some locals have it; also GIRDLE — i.e. *griddle* — STONE) and KIELDERSTONE CLEUGH, which takes us down past the confluence with the CARRY BURN (see the discussion of CARRYCOATS) below UPPER STONY HOLES NT6501 to the SCAUP BURN NT6500. This last reflects MnE *scalp*, with vocalization of the [l] to [u], and meaning "bare piece of rock or stone standing out of water or surrounding vegetation" (they are also to be found on the coast). SCAUP PIKES NY6598 are on the western side (WEST KIELDER MOOR) with SCAUP (a farm) NY6697 below, opposite the confluence with the WHITE KIELDER BURN (below EAST KIELDER MOOR). KIELDER HEAD NY6697 is situated between the two watercourses. To the north-east, we have the intriguingly named OH ME EDGE NY7099 (perhaps some hapless cartographer misheard nearby WOLFMEATH and WOLFMEATH EDGE, see above), and ROB'S PIKES NY6899 (probably ME *pik* "summit", but I have no explanation for the first element). The probable site of Kielder medieval village is below WOODY CRAGS NY6898 (ME *wodi*, so this name might be older than it appears). Below EWE HILL NY6797 (OE *eowu*) and CALLER CLEUGHS NY6797 (OE *cealfru*, ME *calveren* becomes *caller*, so "calves' ravines") there are the remains of more recent habitation in BRAN'S WALLS NY6697, which appear to be the *Brandy Leish's Walls* described in Tomlinson's Guide to Northumberland, p. 228 (*walls* is an element commonly applied to deserted or delapidated buildings).

On the other side of EWE HILL, the RIDGE END BURN rises on OH ME EDGE NY7099, as does OH ME SIKE NY6997, and is overlooked by WETHER LAIR NY7097 (probably OE, ME *weðer* "male sheep" + ME *lair* "soft ground") and MONKSIDE NY6894. If this is the "hillside of the monk(s)" (OE *munuc*) it could indicate the former ownership. The other side of the RIDGE END BURN is fed by SMALLHOPE SIKE, LOWER SMALLHOPE SIKE and SMALEHOPE SIKE, which seem to contain two variants of OE *smæl* "narrow; small". Further down on the other side, the DOVE SIKE NY6897 joins from the north; this may be ME *duve* "dove", but as OE *culfre* also typically appears in this meaning, we should at least entertain the possibility that it is Britt. **dubo-* "black". (BLACK CLEUGH and LITTLE BLACK CLEUGH are two streams which join further down from the south, so there is ample motivation for a colour term.) In the case of nearby HOGSWOOD MOOR NY6795 (probably OE *hogge(s)* + OE *wudu* "wood"), it is anybody's guess which animal the first element referred to: pig, wild boar, or young sheep. The name THREE PIKES NY6695 may refer to the entire ridge extending into NY6594, as there are three peaks, including PIKE KNOWE and GREY'S PIKE. WAINHOPE NY6792 (*Waynhoppe* 1279) is both a farmhouse on the WAINHOPE BURN near where it joins the PLASHETTS BURN and an area of forest. As in so many cases in these upland hills, *hope* (OE *hop*) seems to mean simply "farm", and there is also the regular English word *wain* (OE *wægn* is a different word from modern *waggon*, which seems to be a borrowing from Dutch or German), so, in view of the way in which early English farms specialized in one commodity or another, this may have been a "waggon farm", even though Mawer considers it unlikely. It is just possible that the name continues Britt. **wāgnā-* "moorland, heath, meadow" (Welsh *gwaun*, Old Cornish *guen* etc., which often appears in place-names), and this would not be inappropriate from a topographical point of view. GOWANBURN NY6491 also denotes a house and an area of what is now forest, but there are remains of settlements in the vicinity which may well date back to the Iron Age (i.e. 2,500 to 3,000 years ago). However, the second element of the name is English (OE *burna* "stream") and therefore postdates this archaeology considerably. I am struck by the fact that there is no stream called the Gowan Burn (perhaps the BAKETHIN BURN was formerly so called). It is always possible that the first element is older, although nobody has so far tried to explain it. My immediate thought centres on the name of the divine smith of the Ancient Britons, called *Gofannon / Govannon* in Welsh, which even appears in place-names such as ABERGAVENNY SO2914 (*Gobannion* was the name of the Roman fort by the river). The sound-change from [b] to [v] is quite regular (the old name for the RIVER SEVERN is of course *Sabrina*) and there is sometimes a further change to [w] (Latin *habena* "bridle" becomes *afwyn* in Middle Welsh but *awen* in the modern language), so the development of *Gobannio-* to *Gowan-* is not implausible. Indeed, the absence of a stream called Gowan Burn raises the possibility that the whole name is derived by popular etymology from *Gobannion*. PLASHETTS NY6690 was a third farm building and area combination and, like the PLASHETTS NY9681 near Bavington, may also be derived from Middle French *plaschiet* "little pond" rather than a *plashed* "fence". However, it is now covered by the biggest man-made pond in England (i.e. Kielder Water).

NEEDS HILL NY6690 now forms a peninsula and doubtless reflects OE *nēat* "cattle". To the north, the CHEESE SIKE, which probably has less to do with cheese (OE *cīese*) than gravel (OE *ceosol*, as in CHEESEBURN NZ0971) feeds the PLASHETTS BURN on the northern side of EWE LAIRS NY6591 (ME *lare* "soft ground"). Animal husbandry was not the only activity hereabouts. PITHOUSE CRAGS NY6791 is a modern name (there are several Northumberland locations called PIT HOUSE) but bears witness to the old mining industry in the area, and is perhaps the reason why SONSY RIGG NY6791 is so called (early MnE *sonsy* "bringing good fortune" and *sonse* "luck" from Scots Gaelic *sonas* "good luck"). With PLASHETTS CARRS NY6791, located at the northern end of the crags, we are surely dealing with what the OED calls Old Northumbrian *carr* "rock". CLOVEN CRAG (MnE *cloven* "split, cleft asunder") and WHITE CRAG are at the other end NY6790. From below PIPERS CROSS NY6891, the LONGRIGG SIKE flows down below the long ridge of LONG RIGG NY6891 to the BELLING BURN; the MILLSTONE SIKE below MILLSTONE RIGG and MILLSTONE CRAG NY6892 runs a parallel course further north (presumably the watercourse, ridge and rocks where millstones were to be found). THE BELLING NY6988 is now a peninsula but there used to be a farm on the southern side of the hill by the old course of the river. Perhaps this was at or near the site of an old bell-pit; the more modern Falstone Mine, still worked in the 1990s, was half a mile to the north-east at NY699887. It seems logical to assume that the BELLING BURN was named afterwards and that BELLINGBURN HEAD NY6991 was named after the watercourse. THE LAW NY6788 was another old farm on a hill beside the river, but WIND HILL NY6888 is still well above the water-line. Its etymology would be straightforward (MnE *wind* + *hill*) were it not for the name WHINNY HILL NY6889 just in the next square, which might be one and the same. The adjective *whinny* is recorded as meaning both "exceedingly hard; containing whin or whinstone" and "covered or abounding with whins or furze (gorse) bushes". Take your pick (pun intended).

The area to the east is called HAWKHOPE ['hɔːkəp] and there is still a farm of that name below the Kielder dam NY7188 which incorporates an early-modern bastle house. There may have been a medieval village nearby to which the early forms (e.g. *Haucop* 1325) referred. HIGH HAWKHOPE is just to the north, and HAWKHOPE HILL a little further to the east NY7287 (probably OE *heofoc* "hawk" + *hop* "farm", although a compound of OE *heafod* "head" and OE *copp* "summit" is not out of the question). FALSTONE NY7287 (*Faleston* 1256, *Faustane* 1371) is usually taken to be a compound of OE *fāh* "variegated, multicoloured" + *stān* "stone", but the [l] in FALSTONE is early, so it seems to me that OE *fealo* "dull-coloured, yellow, brown" (which we still have in the MnE adjective *fallow*) is more likely and moreover reflects the colour of the local sandstone. Perhaps the place was named after a particular standing stone. The FALSTONE BURN runs to the south of HOLLOWS HILL NY7288 and rises on GREEN EYES CRAGS NY7388 (could this contain MnE *greeney* "greenfinch"?). The GREEN BURN and the PIT BURN (this is probably the modern word *pit*, but OE *pytt* also meant "pool") flow down from MID FELL NY7489 and POPE'S HILL NY7389 (a gathering-place for recusants or a contraction via haplology from OE *popel* "pebble"?). If we take the road east out of Falstone past MOUNT PLEASANT NY7286, a surprisingly common name which appears to have been especially fashionable in the late 18th and early 19th centuries, we pass STOKOE (already mentioned in connection with

Ridley Stokoe on the other side of the river) and come to DONKLEYWOOD [ˈdʊŋklɪwʊd] NY7486 (*Duncliffe* 1279, *Doncliwod* 1325). This probably means "hill-cliff wood". There may have been a larger village in the Middle Ages, but the evidence is purely documentary (in the so-called *Iter of Wark*). The DONKLEYWOOD BURN runs down from KINGSLEY CRAG NY7487 on (yet another) WHITE HILL NY7488, which is drained by other streams including the MERRY BURN to the west (probably reflecting *mære* "boundary" rather than MnE *merry*) and the RYECLOSE BURN and DEEP SIKE / THORNEY BURN to the east. The farm RYECLOSE NY7486/NY7586 (OE *ryge* "rye", ME *clos* "enclosure") is now surrounded by DONKLEY WOODS. The next stream over to the east is THORNEY BURN, which gives its name to the farms HIGH THORNEYBURN and LOW THORNEYBURN NY7686, and the high ground to the north, THORNEYBURN COMMON NY7687, as well as St Aidan's Church NY7887 (commonly known as Thorneyburn) and the surrounding dwellings. OLD HALL NY7686 is probably the site of the old medieval village of THORNEYBURN (*Thorneybourne* 1325), and this reflects OE *þorniht* "thorny".

To the north beyond THE DODD NY7392, now inappropriately named as it is covered by forest (ME *dod, dodden* "to lop off, cut off, make bare"), EARL'S SEAT NY7192 and HARPER CRAG NY7093, lies EMBLEHOPE MOOR, which derives its name from the farm EMBLEHOPE [ˈɛmləp] NY7494 (*Emelhope* 1325). This is generally thought to contain OE *emel* "caterpillar" (the [b] is non-historical and intrusive as in *thimble*), but which species of caterpillar would have been so prevalent as to merit the name? Perhaps it was someone's nickname. The old pronunciation survives in EMLOPE CRAGS NY7495. BURN GRANGE NY7294 probably takes its name from the confluence of the HARPER BURN NY7093 / BUCKLAKE SIKE NY7094 to the west (the latter reflecting OE *bucca* "male deer" + *lāc* "displaying area", as there are no lakes hereabouts) and the HUNTER'S BURN / BLACKHAG SIKE to the north. It is difficult to see why harpers might provide a motivation for a place-name here, unless it is a personal name, or the popular etymology of another word, such as OE *here-pæþ* "road", or *hor-pol* "muddy pool" (as in HARPOLE SP6960). On the other hand, hunters are easier to visualize in this landscape, for example on HAWK KNOWE NY7294 (OE *heofoc* "hawk") on the slopes of ROUND LAW NY7295 (OE *rymed* "cleared").

Indeed it is the HUNTER'S BURN which joins the SMALLHOPE BURN (if this is OE *hop* in its meaning "narrow valley", then OE *smæl* "narrow" would be especially appropriate) near its confluence with the TARSET BURN NY7397. North of this, it appears to be called the TARN BURN (ME *terne* "small mountain lake"), which rises near STREET HEAD NY7398 and BLACKMAN'S LAW NY7498, and is joined on its journey south by the RUKEN SIKE (probably the same as in ROOKEN KNOWE and ROOKEN EDGE NY8096, but the origin is debated, see above), near OLDMAN KNOWES NY7595, and LONG SIKE NY7496. There is no tarn here nowadays, but there is an area called LOUGH KNOWE just to the west NY7395 (what can this be but ME *louh* "lake"?), so perhaps there was a notable stretch of water here in antiquity. SCAD LAW NY7596 is to the north (of all the *scads* listed in the OED, either no. 5 "salmon fry" or no. 6 "slab of peat, tuft of grass" seems most appropriate here). WETHER LAIR NY7695 (soft ground for male sheep, compare EWE LAIRS NY6591 above) is to the south-east. Nearby ASHY BOG NY7595 is the area to the north of WHITEHEUGH CRAG NY7694 ("white crag crag", the colour of either the rocks or the grass could be meant). English *ashy* usually means "covered with cinders", but this is surely an area covered with ash scrub, especially as ashes from a fire is locally *ass*, as in *ass-pit*. To the south, the area is now forested, so the long narrow hill which gives COMB HILL NY7692 its name is no longer so apparent. The farm at its southern end is also called COMB [kəʊm] NY7690.

There is a collection of bastle houses in the vicinity. First we have SHILLA BASTLE just north of the BLACKLINN BURN "black torrent / pool stream" on SHILLA HILL NY7690, which was probably where corn was winnowed (OE *scyllian* "to separate", or it could reflect either OE *scylf* "peak, crag" or OE *scylfe* "shelf"). Then we have COMB BASTLE by the confluence of the HIGHFIELD BURN and the SMALLHOPE BURN, which is also called CORBIE CASTLE (ME *corby* "raven, carrion crow"). Finally, there is BLACK MIDDENS BASTLE NY7790, an English Heritage property on the eastern bank of what is now certainly the Tarset Burn, which could well reflect ME *midding* "dung, refuse heap", but I am suspicious, as every dwelling would have had a midden, and so offer the suggestion that this is a popular etymology of OE *(ge)mȳþ tūn* "settlement near where the waters meet" (as in the many places called MITTON). Apart from the several streams already mentioned, the BLACK BURN also joins the Tarset Burn near here, flowing down from BLACKBURNHEAD NY7793, over BLACKBURN COMMON, past RIDLEY SHIEL NY7892 ("a summer hut" in a cleared clearing, OE *rydde lēah*, or a summer hut belonging to the Ridleys) and GILLIEHILL CLINTS NY7790. This is ME *clint* "projecting crag"; GILLIE HILL is just above and, if the name reflects early MnE *ghillie* from Scots Gaelic *gille* "lad", it is probably quite modern. The colour black in the names just referred to is no doubt due to peat or coal outcrops. Perhaps there were bell-pits on BELLING RIGG NY7890 to work the coal (there are no workings marked on the map apart from a disused quarry, probably to feed the nearby lime kiln).

SHIPLEY SHIELS NY7789 was probably originally a summer hut in a clearing for sheep (OE *scēap*). Down the road, opposite GLEEDLEE NY7789, a bright clearing (OE *glæd*) or one where kites (OE *glida* "a kite, falcon" survives in Nthb. *glede* "falcon") were to be found, is NEWBIGGIN NY7889, one of the many places simply called "new building", from ME *bigging* "building, house" (ultimately from ON *byggja* "to build"). Across the Tarset Burn is SIDWOOD COTTAGE NY7789, which probably gets its name from the area of woodland to the south-west on the hillside. There are the remains of a Roman-period settlement just upstream, but no onomastic data survive. Downstream, we come to REDHEUGH NY7888, where one naturally thinks of the meaning "red haugh-land" (OE *rēad*, perhaps because of ochre indicative of iron-bearing clay), but it could equally well reflect OE *rydde* "cleared", or even indicate a reedy place (OE *hread* "reed").

Across the burn again, GATEHOUSE NY7888 is probably just a house by the road (like GATEHOUSE OF FLEET and the score of other houses and settlements with this name, reflecting ME *gate*, ultimately ON *gata* "road"). In the bend in the Tarret Burn just to the south we find THE SNEEP (also THE SNEAP) NY7988 (earlier *Snepe*, *Snipe*), which, although it is not listed as a common noun in the OED must have existed meaning "bend in a river" as it is also the name given to a well-known horseshoe bend on the Derwent, nowadays THE SNAPE NZ0449, also SNEEP NT6538, and SNIPE HOUSE NU1508. HEATHERY HALL NY7889, to the left of the road towards RIDLEY SHIEL NY7982, is probably named after its original roofing material, heather thatch; BELLING RIGG NY7890 with its lime kiln is the ridge on the right. Taking the road north-east from Gatehouse leads us towards BURDON SIDE NY8090, where there are no early forms, but comparison with BURDON NZ3851 (*Byrdene* ca. 1050) allows a tentative etymology and a meaning of "cow-shed valley hillside" which sounds strange but suits both the farm and the topography.

SMIDDY WELL RIGG NY8089 is further down the Tarret Burn from Burdon Side and probably meant something like the ridge near with stream where the smithy was (*smiddy* is simply the Scottish and northern dialect equivalent). SUNDAYSIGHT CLEUGH is on the opposite bank of the burn with SUNDAYSIGHT NY8189 (*Sundayheugh* 1325, probably with *hōh* "height") itself on the southern slope of the hill above, hence the sunny hillside (ME *sonny* "sunny"), although a personal name is not out of the question; after all, we have nearby JOHNSIDE NY8088 and JOHNSIDE SIKE NY8188 to the south, as well as LEONARD'S HILL NY8088).

Further up the Tarret Burn, we pass PIT HOUSES NY8191 (recently renamed KEEPER'S COTTAGE, perhaps in pursuit of a more idealized picture of this former semi-industrial landscape) and, as the burn emerges from the forest, catch sight of GIBSHIEL NY8093. As there are no early forms, it is impossible to tell whether this was originally Gibb('s) Shiel (ME *shēle* "hut", of course), or Gibb's Hill. The road itself passes HIGHGREEN MANOR NY8091, built about 1885, so the name is probably not old. However, nearby GIMMERSTONE WOOD, GIMMERSTONE HILL and GIMMERSTONE SIKE which rises between LITTLE DODD and HESLOP CRAG NY7991 probably retain an old settlement name (cf. ME *gymbre* "ewe between first and second shearing", MnE *gimmer*, ultimately from ON *gymbre*). *Heslop* may be a proper name or refer to an original hazel valley (OE *hæsel* + *hop*).

Confusingly from a place-names point of view, the TARRET BURN joins the TARSET BURN near BURNMOUTH NY7988, but with its early forms of *Trivetbourne* (1326) and *Ternetbourne* (1330 — here <n> could have been read for <u> due to scribal error) — *Tarret* immediately becomes a candidate for an early British name. The first element could be Brittonic **trebā* "village, steading" (= Welsh *tref*), which is found all over Southern Scotland in names including TRANENT NT4072 "the village on the stream", TRABROUN NT4674 "village on the hill" and many others. If the 1326 form is most reliable, the second element could be Britt. **bedo*- "grave" (W *bedd*) or **betw*- "birch" (W *bedw*). If we prefer the 1330 form, the second element could be the same as that reflected in the name of the River Nidd in Yorkshire, or Nedd in South Wales. The name of the TARSET BURN, and of TARSET CASTLE NY7985 appear to be English. Although it is unclear what the early forms such as *Tyrsete* 1269 refer to, the castle is a good candidate as it was first recorded in 1267. Mawer reconstructs *Tīra(n)-sǣte* "Tir(a)'s farm or seat", with *Tir(a)* being short for *Tīrw(e)ald* or *Tīrwulf* (eight instances are recorded in PASE). Similar considerations apply to REDMIRE NY7985 as those outlined for Redheugh above, except that clearing a swamp (ME *mir*) is more unusual. SNOW HALL NY7986 is a 19th-century farmhouse, so probably has a modern name. BROWNKNOWE NY7986 is along the contour to the north, and of course means "brown hillock". OE *brūn* is recorded, but, like many colour terms, the word *brown* only becomes frequent in the later medieval period, and was more often used to describe the colour of men's hair, hence the personal name *Bruno* borrowed from early German. All this leads me to think that the place-name is modern in this instance as well. BOUGHTHILL NY7886 on the other side of the burn is named after a sheep-milking fold (ME *bucht*). GREENHAUGH NY7987 (*le Grenehalgh* 1325) is in a green narrow valley as its name implies. In fact, some colour names appear earlier than others, which makes them useful for dating, but OE *grēne* is one of the first, with plenty of examples from the earliest Old English onwards.

The *lane* in LANEHEAD NY7985 probably refers to the long straight piece of road which climbs up from the direction of Bellingham. None of the buildings appear to be older than the 19th century, and we might expect **Loanhead* if the name were anything but modern too. It may have been modernized, however, and the original name preserved in LOAN WELL near the obscurely named BIMMERHILL NY8086. I can only think that WADGE HEAD NY7985 reflects an early MnE or local form of *wedge*, motivated perhaps by a wedge-shaped piece of land (the present house is 18th-19th century). NEWTON and HIGH NEWTON NY7984 are also down by the river and the old railway line, and are likely 19th century too. However, CHARLTON NY8084 (*Charletona* 1279) is older and probably reflects OE *ceorla-tūn* "the farm of the churls or freemen" (*churl* not then having the negative connotations it has since acquired). It naturally gave its name to the CHARLTON BURN, which rises on the high ground to the north and flows, with the RINDS BURN (this is probably *ryn* "very small stream", so another pleonastic formation), through a boggy depression called GREEN SWANGS NY8086 (early MnE *swang* is thought to have the same root as *swamp*). FIELDHEAD NY8086 is a 19th-century shepherd's cottage, and has a correspondingly modern name, although such a dwelling clearly gave its name to nearby SHIEL CRAGS NY8086 (ME *shēle*). There are two other farms up here: CLOSEHILL NY8185, in all likelihood an enclosure (ME *clos*), and BOWESHILL NY8085, maybe so named because the hill was shaped like a bow (OE *boga*), but it is not impossible that it belonged to someone called *Bolla* or the like (as in BOWSDEN NT9941 from *Bollesdene* 1228). Nearby BOE RIGG NY8085 "bow-shaped ridge (?)" might lend credence to the first explanation.

We proceed past LONGHEUGHSHIELDS NY8284 "the summer hut(s) on the long hill" to the north of the road, past THE RIDING NY8284 (OE *hryding* "cleared land" or ME *riding* "way or path made for riders") and THE SHAWS NY8383 "the copses", to BELLINGHAM ['bɛlɪndʒəm] NY8483 (*Bainlingham* ca 1170, *Bel(l)ingham* 1254, *Belingjam*, *Belingeham* 1279) and a discussion of this intriguing name cannot be postponed any longer. The first point to make is that all the forms agree, apart from the earliest spelling (which could therefore be a scribal error). Comparing the name with all the other Northumberland (and southern Scottish) places in *-ingham*, it appears to mean the homestead (OE *hām*) of the family (OE *-inga-*) of someone called *Bella*, *Beola* or the like. It has been shown that names in *-ing(a)ham* form one of the earliest strata of Anglian (i.e. northern Old English) place-names. There is indeed a *Baella* recorded in PASE as a 7th-century monk of Jarrow who told Bede about Cuthbert's miracle of the rafts (and the spelling of whose name may also help account for the diphthong in the aberrant early form). The miracle took place near the mouth of the Tyne, so we cannot place Baella in Bellingham directly, but it is noteworthy that the Anglican church is called St. Cuthbert's. The present building is 13th-century, but like many churches may have replaced an earlier structure; outside is Cuddy's or St Cuthbert's Well, and it is said that St. Cuthbert's body rested in Bellingham *en route* to Durham after the sack of Lindisfarne by the Danes in 875, so a connection between Jarrow and Bellingham is entirely plausible, as is the notion that Baella had a family — marriage was not unusual among early British clerics. Of course, this is only one theory, albeit the most likely to my mind. We should not forget that there are a number of *Bell-* and *Belling-*type names in the area, notably THE BELLING NY6988 and its associated names, and BELLING RIGG NY7890, as well as names such as BELLSHIEL NY8199, which seem to have an association with early mining and the early bell-pit techniques used. One of the early names for Bellshiel is *le Belles* (1370) and this phrase has also been noted in connection with Bellingham (*le Belles* 1326), perhaps in the sense of hills or spoil pits such as those seen on the Woodburn road (although I am making no specific claim regarding these particular heaps), and we should note that the element occurs in various hill names, such as BELL HILL NT8410 and YEAVERING BELL NT9229. So Bellingham could be "the homestead of the people at the pits / hills" or this might have been the source of a popular etymology. We should dispose of one old chestnut: of course, there was a de Bellingham family, but it would seem logical that they took their name from the place rather than vice-versa. Finally, it is not impossible that there is a connection with BELLION NZ0889 [bɛl'aɪən], which seems to be formed in the same way as BULLIONS NZ0353 and FOLLIONS NU0007, i.e. the name of the animal and the dative plural of *lēah* "clearing" which was *lēagum*, and it could then mean "homestead at the hill / pit clearings", but in this case the modern pronunciation is something of a problem.

A view of Carter Pike, NT6904, with its cairn. Northumberland hills called "pikes" usually have a cairn at their summits. Inset: a closer view of the cairn on Carter Pike, NT6904.

On the A68 looking in a southerly direction towards Lumsden Law, NT7205, probably from a pleonastic formation meaning "bare hill hill".

On the A68 looking north towards Redeswire, the modern border between England and Scotland. The name is a compound of *Rede* and *swire* (OE *sweoru* "a gentle depression between two hills"). The site of the Redeswire Fray of 1517 lies just over the border in NT7007.

Looking north towards Whitelee NT7105, "white clearing". The term *white* was commonly used to designate dry open ground, perhaps also the "white" grass so common in this upland landscape (as in the previous photographs, and here in the bottom right-hand corner).

Looking south from Whitelee NT7105 at the upper reaches of the River Rede. For most of its course, Redesdale is a relatively shallow valley.

Catcleugh Farm NT7403 on the edge of Catcleugh Reservoir. Cat Cleuch (spelling is unimportant in this case) is the ravine at the back, presumably where wild cats once lived.

From the A68, Burdhope / Birdhope Farm NY8198 is behind the trees which line the course of the River Rede.

Bellshiel Crag (also known as Burdhope Crag) NY8199. The term crag is often used for rocky outcrops like this.

Looking south on the A68 just north of Rochester village. The choice of the name Brigantium for this now mothballed enterprise was inspired by the tribal name of the Brigantes and represents several real continental names — Bregenz in Austria and Brainçon in France, for instance. However, we are in the territory of the Votadini here, perhaps perserved in the hill-name Woden Law NT7612, just over the border.

The old Manse and Presbyterian chapel in Rochester NY8298, now both in private hands. The chapel was called Birdhopecraig and this name recalls its original position on the high ground to the north. The old chapel dates to 1682 and the "new" building was constructed in 1826.

The farm at Woolaw NY8298 "wolf hill". Despite it having been a sheep farm for most of its history, the name has nothing to do with Modern English *wool*.

Redesdale taken from what is known locally as Saffronside, but if saffron was found here in the past it has laft no obvious traces today. Neither is the name found on the OS map, where we find Saughenside [safənsaɪd] NT7900 "willow hill" instead. *Saffronside* is probably a popular etymology.

Detail from a sketch by Frances Mitford, who writes: "THE House at Birdhope is placed on a rocky Crag above the river Rede. (Reproduction by kind permission of Lord Redesdale.)

The corresponding view from the valley through the eyes of Frances Mitford. (Reproduction by kind permission of Lord Redesdale.)

Frances Mitford's view of "the road to Scotland near Rochester".
(Reproduction by kind permission of Lord Redesdale.)

Redesdale from the Potts Durtrees Road (Blakeman's Law NY8796). Of the several Romano-British settlements in this area there are now few obvious traces. The settlement in the mid-ground left of the picture is Otterburn NY8893.

Frances Mitford's depiction of the Windburn and its "roaring rocky linn".
(Reproduction by kind permission of Lord Redesdale.)

From the Potts Durtrees road, Rattenraw NY8495 is in the middle of the picture on the other side of the river. The old settlement is behind the modern farm and beyond the Rattenraw Burn.

Rattenraw NY8495 taken from Elishaw NY8695.

Upper Redesdale from beyond Potts Durtrees NY8797.

A romantic view of Redesdale near Horsley by Frances Mitford (1863).
(Reproduction by kind permission of Lord Redesdale.)

A romantic view of Redesdale near Horsley by Frances Mitford (1863).
(Reproduction by kind permission of Lord Redesdale.)

Detail of a picture by Frances Mitford of the Roman tombs on Rochester moor (1863). The first is 16 feet six inches in diameter at the foundation; the second is 9'3" square; the third 10' square; the fourth 9'3" square. (Reproduction by kind permission of Lord Redesdale.)

Looking south over the remains of the Roman Officers' tombs by Frances Mitford (1863). (Reproduction by kind permission of Lord Redesdale.)

A stylised view from Todlaw "fox hill" towards the River Rede by Frances Mitford.
(Reproduction by kind permission of Lord Redesdale.)

All that remains of Goodwife Hot NY8778. See text.

Dally Castle NY7784 and the farmhouse below it.

Bent House NY7785 can just be seen in the background, named not after the manner of its construction but after the local coarse grass (OE *beonet*), as reflected also in the name Bennetsfield NY8595.

A misty view of Bellingham NY8483 from the road to Ealingham Rigg. A spoil heap is right centre: could this be one of the bells of Bellingham?

CHAPTER FOUR
4. From Redesdale to Upper Coquetdale

The high ground between Redesdale and Coquetdale is effectively blocked by the 60,000 acres of the Otterburn training area, especially when the red flags are displayed. It is wise to be cautious, or you will be confronted by burly, and ultimately persuasive, military types.

The usual way to travel nowadays is via Elsdon. Leave the A696 outside Otterburn by the B6341. CLOSEHEAD NY9093, presumably the top of the enclosed land (ME *clos*), and OVERACRES NY9093 (*Haveracres* 1538, so most likely meaning "oat fields" with ME *hafyr* "oats" and *acre* in its old sense) are to the north of the road. Beyond that, there is an Iron-Age fort on the appropriately named CAMP HILL NY9093 (there is also one on nearby FAWDON HILL NY8993). On the other side of the road, there is a good view of the valley of the Elsdon Burn and the high ground beyond with THE HAINING (*Hayning* 1304) and HAINING HEAD NY9292 (ME *heining* "enclosed piece of land", ultimately from ON *hegning* from *hegna*). SOPPIT FARM NY9293 (an older *Sokepeth* 1292, lost its unstressed vowel and the [k] was assimilated to the [p] to give *Soppeth* 1323) seems to have been on or near a path over a boggy area (compare OE *soc-bret* "board serving as a footbridge over marshy spots"). It seems that the 19th-century coalmines at Soppit were frequently flooded. ELSDON GATE NY9293, a toll-bar house on the old turnpike (which reveals it as a modern name), marks a turn-off on the ridge to the west of the Elsdon Burn towards THE FOLLY NY9294. There is no country house here, so this is probably not *folly* in its usual sense and but might be a "clump of fir trees on the crest of a hill" (19th-century). It gives its name to FOLLY SIKE, the watercourse which falls into the Elsdon Burn at DUNSHIEL NY9294 (possibly "hill hut"). The track proceeds past FAIRNEYCLEUGH NY9194 "ferny ravine" (OE *fearnig*) and COLWELLHILL NY9194 on the left towards LEIGHTON HILL NY9095. COLWELLHILL is probably named after a cool spring (it is after all near the Iron-Age fort on Camp Hill and the original inhabitants might have chosen the site because it had a good water supply). Although there are no early forms for LEIGHTON HILL, we may compare with GREEN LEIGHTON NZ0291 and surmise that this is another fort or hill belonging to someone called *Leohtwine* or the like (there are over 200 people called *Leofwine* recorded in PASE, so perhaps this is a good candidate). The fact that the road makes a detour round the summit suggests that there may have been a fort here, but an antiquity is not marked on the map. The road continues round the top side of OTTERBURN CAMP NY8995, up to DAVYSHIEL COMMON NY8997 with DAVYSHIEL FARM NY8996 (*Davisel*, *Daveschole* 1344, i.e. Davy's hut) and HOPEHEAD NY8996 on the south-western side (we also have HOPEFOOT NY8895 nearer to Otterburn itself, representing the two ends of an area of land). There is also a turn to the right which takes us to WAINFORDRIGG NY9196, which seems to be a ridge near a ford usable by waggons (OE *wægn*), and HIGH CARRICK NY9296 (CARRICK HEIGHTS NY9096 is to the north-west). LOW CARRICK NY9195 is no longer inhabited. The early forms (e.g. *Cairewik*, *Kayrwik* 1331) support a derivation from Britt. **kagro-* "fort" (Welsh *caer*, etc.) and OE *wīc*, ultimately from Latin *vīcus*, probably also via Brittonic, so the name seems very old indeed. Confusingly, there are two eminences called BLACK HILL almost side-by-side NY9097, NY9297 (if readers will forgive the learned pun).

ELSDON NY9393, which now styles itself "capital of Redesdale", was an important centre in the Middle Ages (old forms include *Elledene* 1226, *Helvesden* 1325) and is usually taken to be the valley of someone with a name such as *Elli* or *Ælf* (the only person with this exact name is a witness to a 10th-century charter, but there are other similar names with <f> which we need to explain the 1325 form). Drive past the church and over the Elsdon Burn before turning left on the B6341. A right turn on the village green will take you up the hill towards WINTER'S GIBBET NY9690, also called WINTER'S STOB (i.e. "stump") and named after a notorious murderer William Winter, hanged in 1791. It shares the site with the Steng medieval wayside cross, of which only the base remains (STENG MOSS NY9691 is just to the north-east, reflecting ME *steng* "pole" probably motivated by the cross). However, if you cross the bridge in Elsdon and then turn right, the road takes you to LANDSHOT NY9493 (probably OE *land* "land" + OE *scēat* "corner, quarter") and LANDSHOT GLEBE (*glebe* "the soil of the earth, a piece of land, especially one assigned to a clergyman as part of his benefice", from Latin *glēba*). LANDSHOT HILL NY9593 is just to the east. This is also the way to HUDSPETH NY9494 (*Hodespeth* 1252), probably the site of a deserted medieval village, the name reflecting OE *pæþ* "path, valley" belonging to someone called Hudd or Huda.

But we are bound for Upper Coquetdale, so turn left after crossing the bridge. Climbing up the long hill out of the village, we pass BOWERSHIELD NY9494 on the right ("dwelling hut") and then BILLSMOOR PARK (hence the PARK BURN). NORTH RIDING and SOUTH RIDING NY9495 (OE *hryding* "clearing") are accessed by a track running parallel to the burn. CARROW HAUGHS is the low ground between two small streams opposite CARROW RIGG NY9496, and these names may contain Britt. **karrowes* "rocks" or **karr-* "rock" + **rī-wā* "slope". BILLSMOORFOOT NY9497 is on the other side of the road just past the car park near JOCK'S CLEUGH NY9496, which is certainly a ravine with a first element seen in a number of local names, maybe just the familiar by-form of *John*. BILLSMOOR indicates a moor or swamp (OE *mōr*) where water-cress was to be found (ME *billere*). DUNN'S FARM NY9396 (the name seems modern, probably that of a family) is accessed by the road behind the car-park. DUNN'S WELL NY9497 is a short distance to the north-east. Beyond is LOANING BURN with the local word for a lane (ME *lon(n)yng(e)* gives *loaning / lonnen*). HEELY DOD NY9298 is a bare hill with a high clearing.

Above that to the north is a cluster of modern and old settlements around THE RAW NY9498, comprising several bastles (post-medieval) and a farmhouse (19th-century). There are no early forms, but the name, not uncommon locally, is generally taken to be OE *rēw, rāw* "row of houses", although this does not conform to our modern idea of what a row of houses should look like (actually, this applies to most of the locations carrying the name). Perhaps they were of timber and have disappeared, or perhaps it was another word entirely. As the site is on a hill, could we be dealing with Brittonic *rī-wā* "slope, hillside", which appears in Cornish names in *-rew* and survives in Welsh *rhiw* "slope"? In this connection, nearby PENCHFORD NY9497, which gives its name to the PENCHFORD BURN is interesting. Again, there are no early forms, so we must consider possible comparisons, of which PENGE TQ3470 is one candidate. This is taken to be Britt. **penno-kaito-* "the end of the wood", and would explain *Penchford* as "the ford at the end of the wood". HIGH SHAW NY9498 could reflect OE *hēah*, but the *high* element usually indicates a later formation (a farm at this site was first recorded in 1673, still late in place-name terms); the second element is OE *scaga* "copse". In HERDLAW NY9498, the second element appears to be OE *hlāw*, ME *lawe* "hill", and the question is which first element suits best. It might be OE *heord* "a herd (of animals)", but this usually results in names in *Hard-*, such as HARDRIDING NY7563 "clearing for herds"; or the older OE *hierde* "keeper of a herd", now usually the second element of compounds such as *shep-herd, cow-herd,* etc.; OE *heorot* "hart" is unlikely, as [t] and [d] are seldom confused.

Beyond that, HEADSHOPE NY9399 (*Heuedshope* n.d., *Headshope* 1618) seems to be an old name, as the earliest form, sadly undated, shows a clear reflex of the <f> seen in OE *hēafod*, but subsequently lost in MnE *head*. Moreover, the genitive points to it being used as a proper name, but there are no such names recorded in PASE. Perhaps it just means "farm of the chief, head(man)". The farm lies between the COMBS BURN to the north and the KEENSHAW BURN to the south. *Combs* may have OE *cambas* "combs, ridges" but OE *cumbas* "narrow valleys" would also be appropriate, the latter a loan from Britt. **kumbo-* "valley" which gives W *cwm* "valley" (as in *Cwm Rhondda*, etc.). The second element of *Keenshaw* is motivated by a copse (OE *scaga*), but the first leaves much room for speculation. OE *cyne-* "royal" (as in KENTON NZ2267, but without the sporadic shortening of the vowel shown there) would give a meaning something like "the copse of the headman", but perhaps this is just too fanciful. At CRAIG NY9399, the bastle is post-medieval, but the name may be much older, as it seems to be straightforwardly Britt. **krakyā* "rock" (= Welsh *craig* "rock, crag", Middle Breton *cragg*). To the north is the PEDLAR'S STONE and DAW'S MOSS NT9300 as well as DAW'S HILL NT9400, the last two reflecting ME *dawe* "jackdaw" (but this word is probably older).

The next farm down the road is GRASSLEES NY9597, i.e. "grass clearings", which gives its name to the GRASSLEES BURN. The plantation on the slope above is called OVENSTONE NY9698. This name could be the farm, that is OE *tūn*, belonging to *Ofa*, as in OVENDEN in West Yorkshire but with genitive *-s*, in which case the name is old; alternatively, the 18th or 19th-century brickworks may have provided the motivation, in which case the name is modern. There was a settlement at PATTENSHIEL KNOWE NY9598, the first element of which is likely to reflect OE *pæþ-tūn* + ME *shēle*, so "the knoll with the hut at the farm by the path". HUMBLE LAW NY9697 is another "maimed, dodded or shaved hill" (cf. OE *hamel*, and HUMBLE HILL NY6481), usually applied to rounded eminences, in contrast to CLOVEN CRAG NY9597, a rocky hill on the other side of CLOVENCRAG SIKE.

The DARDEN BURN flows from DARDEN RIGG NY9896 past MIDGY HA' NY9698 (*Midgy Hall*, 1865 OS map), which was probably originally haugh-land infested with midges rather than a hall. There are two reasons for thinking this. First, it is hard to see why a hall would be midgy, as the flies do not generally venture inside, and second there does not seem to have been a hall anyway, as the three cottages date from the mid 19th century. BROCKLEY PARK NY9697 is the name of the wood to the south, which seems to reflect a clearing where badgers (OE *brocc*) were to be found (the other possibility, OE *brōk* "brook" is unlikely, as this word is not usually found in Northumberland). DARDEN PIKE NY9695 is just to the south-west of DARDEN LOUGH. On the other side of nearby LOUGH CRAG NY9695 a number of small streams join to flow through the ravine called DARDEN CLEUGH. There are no old forms for DARDEN, but it probably contains OE *dēor* "deer" (like DORDON in Warwickshire, or DARLEES NY6764) and either OE *denu* "valley", or OE *dūn* "hill". It is hard to explain DOUGH CRAG NY9795 in terms of MnE *dough* (OE *dāg*) unless it could be a metaphor for bread rising or perhaps a lump of metal (an extension of the meaning found in OE). In view of the number of hills called "Black X", perhaps Britt. **dubo-* "black" (as in Welsh *du* "black" and the name of the River DOVE in Derbyshire, also DOVE CRAG NZ0398, and perhaps DOVE SIKE NY6897) should be considered.

HEPPLEWOODSIDE NY9798 is one of the lost houses of Northumberland and appears both as *Hepplewoodside* (1865 OS map) and as *Hepple Woodhouses* (Tomlinson 1888:344), the confusion probably arising because of the need to distinguish it from HEPPLE WHITEFIELD NY9899, built in the mid 19th century and which probably took its name from WHITEFIELD HILL NY9897. The name WHITEFIELD, which is not infrequent locally (e.g. WHITEFIELD EDGE NU0803 just down the valley), is usually taken to reflect OE *hwīt* "white, light-coloured" rather than OE *hwǣte* "wheat". The lack of unambiguous evidence for wheat cultivation this far north and at these elevations, at the time we assume the place-names were formed, rather supports the conventional view. HAREHAUGH HILL NY9699 is on the northern side of the B6341 and appears to reflect OE *hara* "hare" and OE *halh* "alluvial land" (the farm HAREHAUGH NY9799 lies just beside such a piece of land). However, confusion with OE *hōh* "hill" would not be impossible, given the circumstance that there is an Iron-Age fort on the hill above the farm. Below the hill to the south-east, we find SWINDON NY9799 which is an object lesson in the uncertainties of onomastics. With no old forms to help us, we might first surmise that it contains OE *swīn* "pig, wild boar" and, as it lies in a valley, OE *denu* (which is occasionally realized as *-don*). Then we look across the road to SWINDON HILL with its Bronze-Age cairn and wonder whether the name of the hill might be original and therefore reflect OE *dūn* "hill". Of course, the name might reflect OE *swin* "creek" (hence the SWINDON BURN?) and have nothing to do with pigs. Just to the south and close to the site of another Iron-Age fort (referred to as *Soldier's Fauld* or *Whitefield Camp*) is WITCHY NICK NY9899 (*Witchy Nick*, 1865 OS map), which we could explain as a derivative of OE *wīc*, if, as seems to have been the case (see OED, *wick* n.2), the meaning was generalized from "settlement outside a Roman fort" to "settlement" in general. Alternatively, it could have been a *nick*, in the northern and Scottish sense of "gap or pass between two hills" (and indeed there is just such a gap between Swindon Hill and the hill to the south on which Witchy Nick is located) where wych elms or rowans grew (OE *wice*, ME *wyche* is used to refer to both). I can find no evidence for the name *Witchy Neuk*, which has appeared on the "Keys to the Past" website. The story of the witch Meg o' Meldon resting here on one of her midnight flights has all the qualities of an onomastic tale. On the other side of the Swindon Burn, RIMPSIDE HILL NY9799 probably has the same name as RIMSIDE MOOR NU0806, which has been explained as OE *rima* "border, bank" (MnE *rim*) and ME *sīde* "hillside", although OE *rīm* "rime, hoarfrost" is theoretically possible. The *-p-* is simply a case of stop reinforcement, like the [b] in MnE *thimble* compared to OE *þýmel*, ME *thymelle*, etc.

Beyond Swindon is where the Grasslees Burn falls into the Coquet, which flows down from the its headwaters in the uplands to the north. Up the valley on the western side is HOLYSTONE NT9502 (e.g. *Halistane* 1240, but the early 12th-century Augustinian convent predates this evidence), and HOLYSTONE GRANGE NT9600, which turns out to be an early 19th-century house, so its name will be modern as well. However, there is plenty of evidence of earlier religious activity in the immediate vicinity — for example, the Bronze-Age standing stones called the FIVE KINGS NT9500, of which now only four remain. The stones are to the south of DUES HILL NT9500, which gives its name to the farm DUESHILL NT9601 (the site of DUES HILL deserted medieval village) and associated features. This name has been explained as reflecting OE *dēaw* "dew" and OE *hyll* "hill", but can something as commonplace as dew really have been the motivation? The genitive *-s* suggests that we are dealing either with a personal name such as *Dewi* "David", as in DEWCHURCH SO5331 and SO4831, DEWSHALL SO4832 and possibly in DEWSBURY SE2422 (is a connection with another early saint, St David, too far-fetched?), or it could also be the name of the devil, OE *dēofol*, with loss of intervocalic <f> and vocalization of the [l]. The hill is topped by a cairn and called Beacon Hill by Tomlinson, appearing as THE BEACON NT9500 on the OS maps. The motivation for this relatively common name seems clear, as the hill is visible for miles around, but how old it is must remain a matter of conjecture. Although the word *beacon* "sign, portent" goes back to OE, evidence for the meaning "conspicuous hill commanding a good view of the surrounding country, on which beacons were (or might be) lighted" is comparatively modern (late 16th century).

From Holystone, the course of a Roman road leads back towards the fort at HIGH ROCHESTER / BREMENIUM NY8398. This is clearly how the Romans accessed Coquetdale from Redesdale. There is an Iron-Age fort at CAMPVILLE NT9402, the name of this 18th-century house clearly being motivated by the fort. It may be that the old name of the house was LANTERNSIDE CLEUGH: that is the name of the ravine in which the DOVECRAG BURN flows (named after DOVE CRAG NT9202, cf. DOUGH CRAG NY9795), but it is illogical to suggest that this was where a lantern was hung to warn of Scottish raiders. Who would hang a lantern in a ravine? If there is anything in this story, the lantern (ME *lanter(n)e*, so the name is probably not much older) would have been hung on a hill, such as LANTERNSIDE EDGE NT9301 (OE *ecg* in the sense "escarpment"), which commands a view of the Roman road along which any raiders would surely have progressed. The road runs parallel to the HOLYSTONE BURN over HOLYSTONE COMMON, both secondary names, past NORTH YARDHOPE NT9201 (*Yerdhopp* 1324, SOUTH YARDHOPE is in NT9200). This is one of the cases where OE *hop* may have had its original meaning of "enclosed valley", and would therefore be closely related semantically to OE *geard* "enclosure", *yardhope* providing yet another example of a pleonastic construction, so "enclosure enclosed valley". The course of the road lies with FOULPLAY KNOWE NT8900, the knoll where game birds had their lecking ground (OE *fugel*, with *play* as in COCKPLAY NY8872, DEER PLAY NY8490 and FAIRPLAY NY7650) and CLEMY'S CAIRN NT8800 to the north. DOD HILL NT9100 (compare THE DODD NY7392, etc.), LONG HILL NY9099, GREENWOOD LAW NY8900 are to the south (but the greenwood has long gone). Beyond DUDLEES NT8600, we are almost back in Redesdale. This name may mean "Dudda's clearings", but the first element could reflect ME *dod, dodden* "to make bare, lop off". If, beyond North Yardhope, we had tramped up the LONGTAE BURN (presumably "long toe", but why?), we would have come to WATTY BELL'S CAIRN NT8901, a round cairn of Bronze-Age date, to the east of HIGHSPOON HILL NT9001. With no early forms to help us, there is little hope of deciphering *Watty Bell* or *Highspoon*, beyond saying that *bell* may be "hill" as in YEAVERING BELL etc., and maybe the first element reflects ME *wodi* "woody". There is an OE *spōn* "chip", which has been proposed for SPONDON SK4035 (Derbyshire), so perhaps "hill where shingles were made" for *Highspoon*, but this is clutching at straws.

On the other side of the Coquet, the Roman road strikes out towards CALLALY NU0509 (*Calualea* 1161, so "calves' clearing"), and the Roman fort of *Alauna* near the Devil's Causeway, to which we shall return in due course. In the meantime, we have Upper Coquetdale in our sights. The road passes LADY'S WELL NT9502, which is possibly of Roman date, even though the name is doubtless associated with a much later cult, possibly to do with the convent at Holystone. It is also called NINIAN'S WELL (Ninian was a 6th-century bishop of Whithorn). There is evidence for WOOD HALL NT9503 at least back to the 17th century, and it gives its name to WOODHALL WOOD to the west, as well as, apparently, WOODHALL HAUGH, the alluvial land on the other side of the burn. However, as OE *halh* "haugh-land" is sometimes found as *hall* in modern place-names, it is not inconceivable that the area was originally called something like *Wudhalgh and that this name was subsequently applied to the farm.

HARBOTTLE CRAG NT9202, HARBOTTLE HILLS NT9204, HARBOTTLE LAKE NT9104 (still HARBOTTLE LOUGH in Tomlinson) and HARBOTTLE WOOD NT9303 are all named after the village of HARBOTTLE NT9304 (*Hirbotle* ca 1120), which means either "the dwelling of the army" (from OE *here* "army", a really common word and compositional element in OE). Derivation from OE *hȳr / hȳra* "house of hire / wages / hirelings" or OE *higera*, as in HARLOW HILL NZ0768, "magpie house" is less likely. The earliest form coincides with the building of the castle by the Umfraville family in the early 12th century. The fortification is half encircled by the river which on the western side makes a U-shaped bend called the DEVIL'S ELBOW. The COLDLAW BURN drains the northern slopes of COLD LAW NT9203, understandably a very common hill-name in Northumberland. It joins with the DRAKESTONE BURN to form the BACK BURN (perhaps a local name for this part of the watercourse "behind" the village of Harbottle) and falls into the Coquet near ROCKEY'S HALL NT9404 (which may refer alluvial land as in other cases). The DRAKE STONE NT9204 itself can be seen from the village looking west; it is impressive and invested with enough legend for us to speculate that it might reflect OE *draco* "dragon" (a loan from Latin) rather than the name of the male duck (which seems to be recorded late in English anyway). Bizarrely, the word *drake-stone* is listed in the OED as "a flat stone thrown along the surface of water so as alternately to graze it and rebound in its course" which would be quite a feat with a 30-foot high sandstone erratic.

Beyond WEST WOOD NT9205 (and adjacent squares), presumably so named to distinguish it from HARBOTTLE WOOD NT9303 to the west, and RAM'S HAUGH with RAMSHAUGH PLANTATION NT9205, the road crosses the river near ANGRYHAUGH NT9205 farm. These latter refer again to the land by the river, distinguished in the first case by either wild garlic (OE *hramsa*), or ravens (OE *hræfn, hræfnas* in the plural), or possibly rams (OE *ramm(a)s*, unlikely from a practical point of view), and in the latter case by a word *anger which no longer existed in OE, but seems to have meant "grazing" (compare German *der Anger* "common grazing"). A West Germanic tribe the *Angrivarii* was identified by Tacitus in his description of Germany as living on the lush water-meadows beside the Weser. Moreover, the English word *ing(s)* "water-meadow(s)" is doubtless related, but does not have the *r*-suffix. On the low ground across the river is LOW ALWINTON NT9205, whereas ALWINTON (both pronounced ['aləntən] with loss of [w] as in *Alnwick, Berwick*, etc.) NT9206 (*Alwenton* ca 1240) is on the other side of the RIVER ALWIN on the banks of the HOSEDON BURN. ALWINTON is the enclosure or farm on the Alwin, the early forms of which (e.g. *Alewent* ca 1200) allow us to group it with other British and Continental Celtic river-names ending in *-(w)ent*. There are plenty of examples around the country. There are several rivers with the name DERWENT, also DARWEN and DARENT, all reflecting Britt. **derw-* "oak", so "oak river". In Northumberland itself, we find the BOWENT and BEAUMONT (*fluvium Bolbenda*, ca 1050, possibly Britt. **bow-* "cow river", perhaps IE **bhel-* for which "white" and "swampy" are two meanings which recur). In South Wales, the EWENNY in Glamorgan (*Aventio*, ca 700) is paralleled by the AVANCE in France and AVENZA in Italy, all from *Aventia*, reflecting a root **aw(e)-* meaning "spring, source", which we find in Sanskrit *avatáh* "spring, well", and making **awent-* look as if it is derived from a present participle, so the meaning is perhaps "flowing". The DURANCE in France (*Druentia* in Classical authors), and perhaps also the WENT in West Yorkshire (*Weneta* 1159-60, *Wenet*, ca 1200) could be added to the list. Because there are not too many of these names to work on, analysis is difficult. One problem is deciding whether the ending is *-went* or *-ent*, for example, exacerbated by the observation that almost all the examples have a first element ending in [w]. On the pattern of DERWENT, the ALWIN could mean "water-lily river", reflecting the source of Welsh *alaw* "water-lily" — there is even a river called the ALAW on Anglesey, now dammed upstream from GLANALAW SH3685 to create a reservoir. Whatever the motivation, it must have been a fairly common name, as it also appears in the early forms of the ALWENT BECK in Co. Durham (*Alewent* 1235-6), the ALLEN which falls into the South Tyne near Haydon Bridge (*Alwent* 1275), as well as ALLENDALE (*Alwentedal* 1226), and the ALLOW in Cornwall (if this is the same as Welsh AFON ALAW).

Perhaps now is an appropriate juncture to deal with the name of the River COQUET ['koəkət] (*Cocwud(a)* ca 1050) itself, which rises near the Roman Fort at CHEW GREEN NT7808 and flows through COQUETDALE (*Cokedale* ca 1160) into the North Sea at Amble, off which COQUET ISLAND NU2904 (*Insula Coket*, 1135-54) lies. The name is generally thought to derive from OE *coc* "bird" + *wudu* "wood", so this would mean a wood where game birds were to be found. This is rendered plausible by the fact that Rothbury Forest was a hunting area on both sides of the river, and a parallel name of a forest near Settle (*Cokwode*) exists. It is then supposed that the name of the wood was later transferred to the river, the original name of which has been lost (this is called a back-formation, common enough in place-name development). One suggestion for the original name of the river is *Camel* (the name of the Cornish river CAMEL offers a tempting parallel and the idea is that it survived in the name GAMELSPATH, the name for the continuation of Dere Street which led to the camp at Chew Green); yet another suggestion is *Alwent*, now the chief tributary of the Coquet (and discussed above). All these suggestions ignore the reasonably secure reconstruction of *Coccuveda* as a river-name known from Roman Britain and identified with the Coquet. It represents Britt. **kokko-* "red" (Welsh *coch* "red") + *-wedd*, which could either mean "slope" as in Welsh *llechwedd* "slate slope" or "appearance" (Welsh *gwedd*). The second solution may be supported by the observation that the Coquet is often "filled with red porphyric detritus from the Cheviot" (PNRB 311).

The BARROW BURN appears to rise in an area of boggy land below BARROW HILL NT9004 and flows past BARROW NT9106 to join the Coquet just above Angryhaugh. Even though BARROW is the site of a deserted medieval village, I have not identified any forms to help us distinguish between OE *bearu*, dat. *bearwe* "grove" as in BARROW in Gloucestershire, and OE *beorg* "hill, mound" as in BARROW in Somerset. As there is an area marked BARROW SCAR just above the Coquet NT9006, it is possible that the ridge as a whole, which resembles a burial mound from some vantage points, including the part now called CALF LEE NT9005, originally bore the name and provided motivation for the name of the burn and the village. The boggy area mentioned above is fed by the WHITELEE SIKE "white clearing stream",

Cheviot Hills and Dales

which rises on WHITELEE KNOWE "white clearing knoll" NT8803; the WILKWOOD BURN, which rises on the high ground to the east of BUSHMAN'S CRAG NT8403 and is clearly paired with WEST WILKWOOD NT8703 and EAST WILKWOOD NT8902 (*Wilkewde* ca 1230, probably the name *Willoc*, diminutive of *Willa*); the SMALL BURN (likely in the sense "narrow") and RAMSEY'S BURN (this also looks like a personal name), which rise on the eastern slopes of BLUESTONE EDGE NT8602 (*bluestone* was probably obtained at the quarry here). LINSHIELS LAKE, between LONG CRAG NT8904 and SLIPPERY CRAGS NT8804 (surely both modern names), clearly belongs with LINSHIELS NT8906 on the southern bank of the Coquet. There have been attempts to explain the name as "lime-tree huts" (OE *lind(e)*), and the objection to this is not that the trees would not grow there (I am writing this looking at lime trees growing at a similar elevation in a more exposed position), but that it is on the basis of one of the early forms (*Lyndesele* 1314), whereas the others (e.g. *Lynsheles* 1292, *Linesheles* 1324) suggest OE *hlynn* "torrent, waterfall, pool", and indeed there is one just upstream from NICHOL'S POOL NT8706. In addition, it is not impossible that the lake provided the motivation for the name of the farm, and if this is so, we could have OE *līn* "flax" as a possible first element. LINBRIGGS NT8906 is on the other side of the river and is reached by a bridge before the road crosses the river yet again by another bridge, hence *briggs*, and the -*s* plural which is late (OE *brycg* was feminine and therefore had a plural *brycge*).

The site of a medieval village is in NT8806 (usually identified with QUICKENING COTE, but OLD QUICKENING COTE is further up the burn NT8706). The name is either Nthb. *quicken(s)* "weeds, couch grass" or more likely *quicken(tree)* "mountain ash, rowan" with OE *cot(e)* "cottage"). It is on the RIDLEES BURN, reflecting in its turn OE *rydde lēahas* "cleared clearings". We should note that WITCH CRAGS NT8705 probably also contains the name of the rowan-tree, i.e. Nthb. *witch(wood)*. The burn rises on the watershed between Redesdale and Coquetdale just beyond RIDLEES HOPE NT8206 "cleared clearing valley or piece of land") and is fed by other small streams. These include RUNNERS BURN (Nthb. *runner* "small stream", this is OED meaning 28a!!) below WOOLBIST LAW NT8207 (can this reflect OE *wulfes fīst* "toadstool"?). Then we have the SOUTHHOPE BURN below BROWN LAW NT8305 (both reasonably self-explanatory as "south valley stream" and "brown hill"). GRAHAM'S CLEUGH (a ravine and what seems like a personal name) crosses the so-called RIDLEES ROAD (a drove road?) on a dog-leg bend. The intriguingly-named PUDDING BURN flows down between SCALD LAW NT8307 (OE *sceald* "shallow, low"?) and GIMMER KNOWE NT8407 (ME *gymbyre* "ewe between the first and second shearing"). CRIGDON HILL NT8605 seems to be another pleonastic formation with Britt. **krūko-* "hill" + OE *dūn* "hill" plus later *hill*, so the meaning is "hill hill hill"! On the northern side of the RIDLEES BURN, the hills worth mentioning include ROOKLING LAW NT8506 which looks down on CORBY LINN NT8506. The word *rookling* is recorded as "a young rook", but perhaps the original name reflected Nthb. *rouky* "misty" and was remodelled by popular etymology because of the proximity of CORBY LINN, which seems to reflect ME *corby* "rook, crow" and OE *hlynn* "waterfall". The burn does not seem to have a name recorded on the map, but it rises on the southern slopes of FALLOW KNOWES NT8507 (OE *fealo* in the sense of "dull-coloured, yellow"). The slopes of CLIFTON RIGG NT8606 above the Ridlees Burn are perhaps steep enough to justify the name (i.e. "cliff hill ridge").

INNER HILL NT8707, for which I have no explanation apart from the obvious, looks down on SHILLMOOR NT8807, one early form of which (*Schouelmore* 1292) has been taken to indicate a moor (OE *mor* "heath, moor; bog, etc.") belonging to someone called *Scufel* (but note this is a hypothetical reconstruction). Perhaps we are dealing with a metathesized form of OE *scylfe* "shelf, river-bank" instead. Here, the road crosses the river yet again and runs just north of east under yet another INNER HILL NT8708. A stretch near the road is marked KATESHAW CRAG NT8707, which, like KATESHAW HILL NY6586, might indicate a copse where wild cats were found. Names such as THROUGHAM in Gloucestershire are usually taken to reflect OE *þrūh* "water pipe, conduit" and to refer to a deep valley, and the Coquet valley below THROUGH HILL NT8607 is certainly deep and steep-sided enough to justify the comparison. PATHLAW SIKE skirts its northern slopes, running down from PATH LAW NT8607 (perhaps PATHLOW in Warwickshire is a parallel, in which case "way hill" or even "wayfarer's hill" is a possibility) and LONG HILL NT8507, which is indeed a hill with an elongated ridge.

CROFT SIKE (OE *croft* "enclosed land used for tillage or pasture") falls into the Coquet beside BYGATE HALL COTTAGES NT8608, which clearly belong with BYGATE HALL NT8508 (first recorded in 1712, now used for storage). If the name contains OE *byge* "bend in a river" or OE *byht* "bight, bend of a stream", we might suppose that the settlement by the Coquet, where there is a bend, came first. The second element could signify a gate (i.e. the northern form of OE *geat*), or a road (ME *gate*, ultimately from ON *gata*), in which case the proximity to PATH LAW NT8607 may be significant. The hall looks down on the DUMBHOPE BURN, a tributary of the DEERBUSH BURN (which rises north of DEERBUSH HILL NT8308) to flow through a steep-sided valley marked DUMB HOPE NT8509 below DUMBHOPE LAW NT8508. OE *dumb* meant "silent" as well as "bereft of speech", so perhaps this was a particularly silent enclosed valley. Just possibly we are dealing with Britt. *donnos "dark" (cf. Welsh *dwn* "dark red, brown, swarthy", perhaps Gaulish *Donnos*, although the meaning of this is uncertain), which would have a semantic parallel in FALLOW KNOWES NT8507 and maybe the BLACK LINN waterfall NT8409 ("black waterfall") a little way upstream. If we include BLACK KNOWE NT8209 just to the west of Deerbush Hill, there is a notable cluster of names denoting dark colours in this valley. DEERBUSH HILL, with HOG KNOWE NT8308 just to the south, seem to refer to game (OE *dēor* "wild animal, deer", OE *hogg* "wild boar, hog"), although wildlife must be as rare as any thickets (OE *bysc*) as this is an impact area for the army range.

Upstream from the DUMBHOPE BURN, the road continues through a deep valley with SHILLHOPE LAW NT8709 (a connection with

Cheviot Hills and Dales

SHILLMOOR NT8807 seems likely, but the ultimate etymology is just as uncertain) rising on the eastern side. BARROWBURN NT8610 takes its name from the stream which falls into the Coquet just downstream, which in turn takes its name from BARROW LAW NT8611, which contrary to our expectations of a prehistoric burial place turns out to be "briary hill" on the basis of the old forms *Brerylawe* 1304 and *Brerilawe* 1307 (ME *brere* "briar"). The name must have been changed later by popular etymology. Further upstream, HEPDEN BURN is the name marked on the map (*Hepden burne* 1153-95), probably reflecting the valley where brambles or hips (OE *hēope*) were to be found. It is separated from the Usway Burn by MIDDLE HILL NT8712, and rises on the eastern slopes of LITTLE WARD LAW NT8614, which is slightly higher than WARD LAW itself NT8613 (from OE *weard* "watch") where there are the remains of an Iron-Age village and fort. Back near the Coquet, LOUNGES KNOWE NT8610 (*Loundering Know*, 1663) may be connected with Nthb. *lounder* "to beat, cudgel", or this may be a popular etymology of an obscure name. By comparison, WINDYHAUGH NT8610 seems transparent, but the old form (*Wyndihege* ca 1200) suggests ME *hege* "enclosure".

Back on the road, we now follow the Coquet in a north-westerly direction before meeting the confluence of the ROWHOPE BURN. This place NT8511 is known as TROWS ROAD END, or SLIME / SLYME FOOT, but neither of these names appears on the maps. The first is transparent enough and the second probably reflects OE *slīm* "muddy place" (as in SLIMBRIDGE in Gloucestershire) with the common northern element *foot*. A private road eventually leads up to ROWHOPE NT8512 (probably OE *rūh* "uneven, uncultivated"), which gives its name to the burn, and then along the TROWS BURN towards the deserted farm of (THE) TROWS NT8512 (*Wytetrowes* 1197) which seems to have been motivated by the pools in the burn (OE *trog*, etc., the word can mean a trough-like pool or depression). An alternative explanation is trolls used for fishing with vocalization of the [l], but we might expect [l] in the early forms if that were the case. The track continues towards the border ridge and WINDY GYLE NT8515, while the burn flows to the west of TROWS LAW NT8513 from its sources. These include ROUTIN WELL NT8514, which could, like ROUGHTING LINN NT9836, contain ME *routand* "roaring", said to be the first element of RAWTENSTALL in Lancashire. ROUTINWELL STRAND is a small stream (ME *strand* "rivulet" of obscure origin) which joins with INNER STRAND and a stream which flows through LOFT CLEUGH to form the main watercourse. LOFT HILL NT8413 is just to the south (probably containing OE *lyft*, *loft* "air, wind, sky"). For those wishing to walk up to the border ridge, THE STREET, an old drove road, joins the modern road where the ROWHOPE BURN crosses it. HINDSIDE KNOWE NT8411 may contain OE *hind* "female deer" or ME *hind* "raspberry", but ME *sīde* and *knoll*, both meaning "hill(side)", are a sure sign of another pleonastic formation. Further uphill, BOUGHT LAW NT8412 and SWINESIDE LAW NT8313 bear witness to the course of the drove road (ME *bought* "sheep-milking fold", OE *swīn* "pig, wild boar"); BEEF STAND NT8213 and BEEFSTAND HILL NT8114 on the other side of the

CARCROFT BURN (see below) possibly relate to the practice of transhumance in the uplands. On the other hand, BLACK BRAES NT8314 and MOZIE LAW NT8315 (OE *mos* "marsh") indicate that it is very peaty and boggy up here. The summit of WINDY GYLE NT8515 has a prehistoric cairn subsequently associated with Lord Francis Russell who was killed nearby in 1585 and is therefore called RUSSELL'S CAIRN. It is typically very windy on the summit, so we can immediately account for the first element, and the second may be Nthb. *gowl, gool, gyle* "a hollow passage or pass between hills" as reported in Heslop (p. 337). If this is correct, we are left wondering which hollow passage is being referred to; perhaps it is the valley of the Gyle Burn just over the border. There are a number of similar rounded grassy hills in Scotland called GEAL (e.g. GEAL CHARN NJ2810) and this is usually taken to be Old Irish *gel* "white" and to refer to late-lying snow. The equivalent in Welsh is *gell* "yellow, brown" (and cognate with English *yellow*, etc.), but there is no evidence that W *gell* ever had the meaning "white" and the phonological development is problematic. However, there is a Welsh word *cil* "back, nape of the neck; retreat, recess, nook etc." which seems to account for the first element of GILCRUX ['gɪlkruːz] in Cumbria and CULCETH ['kʊltʃəθ] in Lancashire, which goes back to *$k\bar{u}los$ (cf. OC *chil* glossing *cervix*). In early Brittonic, [uː] regularly becomes [iː] which would then become English [aɪ] via the Great Vowel Shift; initial [g] for [k] is as we see in the Cumbrian form and is either due to initial mutation (as in GILFACH ST1598 in Glamorgan) or to analogy with some other word (such as Nthb. *gowl*, etc.). So the name might mean "windy nook, retreat", perhaps even with reference to the earthworks near the summit.

After these dizzy heights, if we continue up the course of the Coquet, we come to CARSHOPE NT8411. There are no old forms as far as I'm aware, but one commentator has suggested a valley farmed by a churl (i.e. OE genitive *ceorles*). This sound change is very unusual (especially with all the local Charltons) and OE *cærse* "cress" or Nthb. *carse* "marsh" are likely sources. In fact, nearby CARLCROFT NT8311 may also contain the word for cress, like CARLSWALL in Gloucestershire with late intrusive -*l*- (*Crasowel* DB, *Kersewell* 1220) rather than a Scandinavianized form of OE *ceorl* (and OE *croft* "enclosed land used for tillage or pasture" of course). CARLCROFT HILL NT8312 is just to the north. The CARSHOPE PLANTATION south of the river (a modern name) almost hides BELL HILL NT8410, which provides yet more evidence for OE *belle* meaning "hill"; it certainly does hide WHAR MOOR NT8311, which may, like WHARMLEY NY8866, have been where millstones were found (OE *cweorn* "millstone, quern"). Only closer inspection will tell.

BLINDBURN NT8210 is located near the confluence of the BLIND BURN with the Coquet. There are no early forms, but the name is quite common (there are at least two others in Northumberland, not to mention the ones in Scotland) and probably contains OE *blind* in the sense "dark, obscure, secret"; we might also note that a blind creek is one that sometimes dries out, even though the recorded example is very late. BROADSIDE LAW NT8211 (probably "broad hillside hill") is to the east of the burn, and YEARNING LAW NT8811 is to the west, with

YEARNING HALL NT8112, now a ruin, on its northern slopes. ME *yearning* is a word for rennet and *yearning grass* was used to curdle the milk to make cheese. LAMB HILL NT8113 is at the head of the valley on the border ridge, and there is no reason to think that this does not reflect OE *lam(b)* etc., especially as WEDDER HILL NT7911 (OE, ME *weðer* "castrated male sheep") flanks it to the south-west and BEEFSTAND HILL NT8214 to the north-west. We have late ME *beef* "flesh of a bovine", but the sense "beef animal" is not otherwise recorded until the 19th century, suggesting a late name here.

The RENNIES BURN (redolent of indigestion remedies, but more probably named after *reans* "cultivation strips") rises on the northern side of WEDDER HILL and joins the BUCKHAM'S WALLS BURN to the east of the settlement NT7911. Ruined buildings are often called *walls*, and *Buckham's* seems like a personal name, although one always has to bear in mind that OE *buc* and *bucca* could signify both wild deer and goats (of which there are still plenty in this area), and the name may be another popular etymology. Near Buckham's Walls, FOUL WHASLE is a stream which flows around the south-eastern slopes of PETE'S SHANK NT7910. The word *shank* was often used in early MnE in Scotland and the north to mean "the projecting part of a hill, narrow ridge", and *Pete* may just signify *peat*, used for fuel, walls, etc. Its use in place-names seems to antedate its general currency in the language and it may even be a Brittonic loan word. OE *twisla* meant "fork of a river", and I presume this is the source of *whasle*, as Scottish forms include *twustle*, making the occurrence of [a] rather than [i] less unlikely, and the initial *t-* is often lost in place-names (usually after a word ending in *-t*), the first element being OE *fūl* "foul, unclean". This element recurs in FULHOPE NT8110 "foul, unclean piece of land / valley", where the FULHOPE BURN joins the Coquet. FULHOPE EDGE NT8209 (ME *egge* etc. "ridge") is the long hill to the south-east. Between this and SADDLER'S KNOWE NT8109 is SADDLER'S SLACK NT8109 (ME *slack* "small shallow valley" is from ON *slakki* with the same meaning according to the OED; OE *sadol* "saddle" forms the basis of several saddle-shaped features, e.g. SADDLEWORTH SD9805, if we are not dealing with a personal name here). THIRL MOOR NT8008 is on the other side of the Fulhope Burn and probably contains OE *þyrel* "gap", perhaps referring to the narrow valley of the Coquet to the north, in which MAKENDON NT8009 is situated. In view of the fact that this name has long been associated with the Roman camps at Chew Green, it is tempting to interpret *-don* as Britt. **dūn(om)* "fort". The first element has been taken to be a personal name *Macca*, which could have had *-n* in the genitive, as it was a so-called weak noun. For a solution which does not rely on a personal name, we may compare British Latin **mācēria* (= Welsh *magwyr* "wall, rampart", Old Breton *Macoer* glossing *vallum*) which appears in several place-names, e.g. (ASHTON IN) MAKERFIELD in Lancashire or MAKER in Cornwall, the *-n-* perhaps being a relic of the definite article, yielding a possible and not inappropriate meaning "ramparts of the fort".

We could take BROWNHART LAW NT7809 at face value and surmise "the hill of the brown female deer", but the word *brown* as a colour is relatively modern (remember, colour terms develop late), so a popular etymology might be preferable. Brittonic *brunnyo- meant "hill" (Welsh *bryn*, etc.) and this gets us almost half way, and the second element could be *ardu-* "high" (Welsh *hardd* "beautiful" seems to be a secondary development, the older meaning being preserved in Old Irish *ard* "high" and the Gaulish name of the Ardennes, recorded as *Arduenna [silva]*), or *altā* "slope" (which also appears in place-names, as we have seen in the case of OTTERCOPS NY9589). The second element could even have been *artos* "bear" (Welsh *arth*, C *ors*), as the brown bear was hunted to extinction after the Romans arrived but before the English invasions of the post-Roman period (so it is not surprising that there are no English place-names containing the word OE *bera* but several containing *arth* in Wales, e.g. GLYNARTHEN SN3148, ABERARTH SN4763). Whatever the solution to this question, the river below HARDEN EDGE NT7807 is the last stretch of the Coquet before the source, which is marked as being in Scotland (COQUET HEAD NT7808), but it is impossible to be sure exactly where it is. If HARDEN EDGE is to be compared with HARDEN SE0838 (*Hareden* 1166), it is named after the valley frequented by hares (OE *hara* "hare"); if it is to compared with HARDEN SK0101 (*Haworthyn* 14c) or HAWARDEN SJ3165 (*Haordine* DB) then a high piece of land or enclosure is the motivation (OE *hēa* + OE *worþign*). The border between England and Scotland to the west of Chew Green camp is appropriately marked by the MARCH SIKE, probably reflecting ME *marche* "border".

CHAPTER FIVE
5. From Hepple to the Devil's Causeway

This chapter is for those who choose not to take the road up to the head of Coquetdale, but instead press on towards Rothbury. Drivers cross the Coquet on the B6341 and arrive at HEPPLE NT9800, which appears to reflect OE *hēope* "bramble / hip" and either OE *dāl* "narrow valley" or OE *halh* "land beside the river". This goes some way to explaining the alternation of <d> and <h> in the early forms (*Hepedal* 1199, *Hyphehal* 1229). The medieval village was probably on the southern side of the river, where there is indeed haugh-land, although there are also remains of possibly medieval dwellings on KIRK HILL NT9700 (the name is probably motivated by Hepple Chapel, which was in ruins by 1725). This high land is between WEST HEPPLE NT9700 and THE SCROGGS NT9700 (ME *scroge*, "brushwood, underwood"). EAST HEPPLE NT9800 is just to the north of a small hill called THE CRUTCH NT9800, which may indicate the presence of a cross as in CRUTCH HILL NU1728, but there is no trace of this now, so the name might simply reflect Britt. **krūko-* "hill" (W *crug*, etc.). WREIGHILL NT9701 ['riːhil] and WREIGHILL PIKE NT9802 are to the north. The early forms (*Werghill* 1292, *Werihill* 13c) point to OE *wearg* "criminal", so the hill may have had a gallows or other place of execution (the word has the same root as German *würgen* "to strangle"). The WREIGH BURN is therefore "felons' burn", so criminals may have been drowned there. Perhaps Coquetdale will eventually reveal its bog bodies. Below this grim hill is LOW FARNHAM NT9702 and HIGH FARNHAM NT9602 (*Thirnum* 1242, 1307, *Thernhamme* 1343), which either reflects OE *þyrnum* "by the thornbushes" with the dative plural functioning as a locative, or OE *þyrne-hām* "thorn home(stead)". This may come as a surprise to those who think that the change of [θ] to [f] is a modern phenomenon: perhaps our ancestors also *fought* when they *thought*.

SHARPERTON NT9503 (*Scharberton* 1242) is close to SHARPERTON EDGE NT9604, a steep hill, so the name may reflect *scearpa beorg* "steep hill", or the name was originally *scearda beorg* "notched or gapped hill" (there is a phrase *to ðæm sceardan beorge* "at the notched hill" in the *Cartularium Saxonicum*) and the name would have been changed to *Sharp-* by popular etymology. On the southern slopes of the hill is CHARITY HALL NT9604 (ME *carited* usually indicates a hospital or similar charitable institution) and on the northern slopes is the course of the Roman road, which runs roughly eastward over the FOXTON BURN. To the north is the site of the medieval village of FOXTON NT9605, the early forms of which (e.g. *Foxden* 1324) point to a meaning "fox valley". On the other side of the burn is BURRADON NT9806 (*Burhedon* 1242), probably OE *burg* "fort" + *dūn* "hill / hillfort", but no trace of the defences now remains. It is typical of such places to be located near Roman roads. BURRADON MAINS NT9606 (short for *demesnes*), LOW BURRADON NT9705, and BURRADON WINDYSIDE NT9805 are secondary names where the motivation for the additional elements is fairly clear. Further east along the road, TREWHITT ['trʊfit] (*Tirwit* 1150, *Tyrewyt* 1229) gives rise to another group of names: HIGH TREWHITT NU0005, LOW TREWHITT NU0004, TREWHITT HALL

NU0006 and TREWHITT STEADS NU0006 as well as TREWHITT MOOR NT9804. The name is somewhat perplexing, and some of the derivations offered (e.g. from Nthb. *tirwit* "lapwing") decidedly odd. Moreover, the widespread use of the name makes it difficult to identify the location originally referred to. However, it is worth noting that Britt. **trebā* "settlement" (W *tref* "town" etc.) is sometimes realized as *Tir-* (as in one early spelling of TRABBOCH in Ayrshire, namely *Tirbeth*). As for the second element, one possibility is Britt. **bedo-* "grave" (with <w> for <b> in an attempt to indicate the initial mutation?), which could have been motivated by the Bronze-Age cist (marked as a tumulus on the OS map), so the Ayrshire town may even have the same second element as well.

Back on the north bank of the Coquet, CAISTRON NT9901 (*Cers* ca 1160, *Kerstirn* 1202) is probably either simply ME *kers* (Nthb. *carse* "marsh"), perhaps with OE *þyrne* "thorn" added. An enclosure (originally topped with thorns?) lies to the east of the old village and runs down to the Coquet, but Tomlinson's idea that "Cester-ton" provided the motivation for Caistron is untenable. BICKERTON NT9900 (*Bykerton* 1245) is on the other side of the river and was probably the location of a honey-farm (OE *bēocere* "bee-keeper"), sheltered from the prevailing wind by COTE HILL NT9900 (ME *cote* "cottage"), BLACK BRAE NT9800 and BICKERTON KNOWE NY9999. The eminence to the south (440 m.) is TOSSON HILL NZ0098, which seems to reflect OE *tōt-stān* "look-out stone", one of the most westerly of the Simonside Hills. Moving along this group towards the west, RAVENS HEUGH NZ0198 "ravens height" rises to 422 metres, and the top of SIMONSIDE NZ0298 itself is 429 m. The old forms (e.g. *Simonseth* 1273, *Simundessete* 1278) point to the seat (OE *sēte*) or hill (ME *sīde*) of Sigemund (who also features in SIMONBURN NY8873). There are four people called Sigemund and one Sigmund listed in PASE and, although none of them seem to fit our location, the name was widely used in English (cf. SYMONDS YAT SO5516 in Herefordshire) and is not strong evidence for Scandinavian settlement. These hills are one of the more salient features of the Northumberland landscape and are visible from a surprisingly wide area. Between Simonside and Ravens Heugh is MAIN STONE NZ0298. This could reflect ME *maine* "of great size or bulk", but we should note that Britt. **magno-* "stone, standing stone" (W *maen*, B *men*, as in *menhir* etc.) could yield this element as well, as in TREMAIN in Cardiganshire. OLD STELL CRAG NZ0398 (the old *stell* "enclosure for stock" might refer to the cairn) is further to the east, and DOVE CRAG NZ0398 (see the discussion of DOUGH CRAG NY9795 above) further over still. On the northern slopes, GREAT TOSSON NU0200 sits under BURGH HILL NU0200 with its Iron-Age fort (OE *burg* "fort"), contrasting with LITTLE TOSSON NU0101 which overlooks the gravel workings at Caistron. WOLFERSHIEL NU0100 is a small farm overlooking one fork of the Chesterhope Burn. There is a noun *wolfer/wolver* "wolf-hunter", but the examples in the OED are late, later indeed than the place-name (*Wolfershield* on the 1865 OS map). CHESTER HOPE NZ0299 (marks an area further up the hill and may have been motivated by the remains of the small univallate hillfort in the immediate vicinity. Nearer the river, RYEHILL NU0201 reflects a common place-name, and ALLERDENE NU0201 is not unique either, although there are no old forms associated with it to help us distinguish between OE *alor* "alder" and a personal name such as *Aelfhere* which underlies the other ALLERDEAN NT9847.

On the other side of the river, we pass through FLOTTERTON NT9902 (*Flotweyton* ca 1160) which is explained by OE *flot-weg*, a path made from wooden floats, known in Ireland as a *togher*, and a technique widely used in prehistoric Europe to carry walkways over boggy ground. WARTON NU0002 (*Warton*, ca 1250) implies OE *weard* "watch, lookout". The site of the medieval village was in the field to the south of the present buildings, overlooking the Warton Burn. SNITTER NU0203 (*Snitere* 1175) is situated on a narrow ridge between the Wreigh Burn and the Back Burn. The name could be connected with ME *sniter* "to snow" and reflect the exposed location, or with OE *snīte* "snipe" and recall local game birds, but the formation is unclear to me (and it has foxed Mawer and Ekwall as well). SNITTER WINDYSIDE NU0104 is so called because it is in Snitter civil parish: *Windyside* just means "windy hillside". SNITTER MILL NU0302 is on the BACK BURN (also called the LORBOTTLE BURN in Tomlinson p. 352), just above THROPTON NU0302. LORBOTTLE NU0306 (*Leuerboda* 1176, *Leuerbotle* 1178) is a farm near the headwaters of the burn, and the site of the medieval village, which could contain OE *botl* "dwelling" and a personal name, for which *Leofhere* is one plausible suggestion. LORBOTTLE WESTSTEADS NU0207 is (naturally) just to the west near the course of the Roman road; LORBOTTLE HALL NU0408 is 18th-century. The meaning of WHITTLE NU0204 depends on the derivation of the first element (the second is OE *hyll* "hill"), which could be *wīt* "punishment" (this might tie in with the Wreigh Burn, see above), or *hwīt* "white, fair". The present buildings are early 19th-century. CARTINGTON NU0304 (*Cretenden* 1220, *Kertindun* 1236) medieval village was located to the west of the castle. The early forms suggest that the second element of the name either meant "valley" (perhaps that of the SPOUT BURN which rises on the slopes of SPOUT HILL NU0404, *spout* possibly referring to the waterfall) or "hill" (CARTINGTON HILL NU0405, would then be a pleonastic formation, is to the north-west). For the first element, we are either dealing with a personal name (*Crettinga* "Cretta's family" is one suggestion) or a derivation of OE *ceart* "rough common". Back towards the river, it is tempting to connect GLITTERINGSTONE NU0303 with the Bronze-Age cairn just to the east. If you're a fan of glitter rock, this is clearly a monument before its time. On a more serious note, as these stones probably do not glitter more than any others, W *cludair* "heap (of stones, wood)", plural *cludeirion*, is found in the hill-names GLYDER FAWR SH6457 and GLYDER FACH SH6558: <c> for <g> is explained by initial mutation; [u] regularly becomes [i]; weakening of unstressed vowels accounts for the other changes and the old plural ending is reinterpreted as a present participle. It would then mean "heap of stone stones", exemplifying the pleonastic construction so frequent in place-names. LYNNHOLM NU0302 may well contain OE *līn* "flax" with OE *holm* meaning either "low lying ground by the river" or "hill" (the latter is much rarer). On the banks of the Back Burn / Lorbottle Burn are the Chirnells (BLACK CHIRNELLS NU0303, BLUE CHIRNELLS NU0303 and RED CHIRNELLS NU0302), clearly connected with CHIRNELLS MOOR, which Dixon, *Upper Coquetdale*, p. 302, identifies with *Childerlund* 1178, *Chirlund* 1167, *Chirland* ca 1250 (CHIRNELLS NU0322302866 is no longer marked). There seems to be little motivation for OE *cildra* "children" to be the first element, and elsewhere *Chir-* points to a church (as in CHIRTON NZ3468), so *Childer-* may be an early popular etymology (ON *lundr* "grove" has been suggested for the second element). CHIRNELLS MOOR is no longer marked on modern maps, but was in NU0403 according to the 1850 OS map, to the north of which is the site of the Chapel of St Ellen, with an associated well, and we do have an indigenous **landa* (as in *Vindolanda*, also W *llan* "church"), which would explain the 13th-century form *Chirland* as yet another pleonastic construction (with OE *cirice* "church"). If OE *cirice* explains the first element of Chirnells as well, perhaps the second element is derived from the name of St Ellen, either via metathesis or the loss of some intervening element, or by analogy with another word entirely, perhaps ME *carnale* "charnel house" from Latin *carnarium*. This is all very uncertain territory.

THROPTON NU0302 (*Tropton* 1177) reflects OE *þorp / þrop* "dwelling, village", so need not be an indication of Scandinavian settlement, and probably meant "village farm". Towards Rothbury is the wonderfully-named village of PONDICHERRY NU0401, which on the face of it is the anglicized name of one of the principal French settlements of 17th-century India (now *Pudcherry*, "new village" in Tamil). It is possible that an owner named the village after the place in Tamil Nadu, but I have not uncovered any evidence for this. I cannot help noticing the two Iron-Age forts on opposite hills on the north side of the river, one nowadays called West Hills and the other called Old Rothbury, and, if there was a bridge over the river (perhaps the modern footbridge continues an old right of way), one could imagine a place called "bridge of the forts" containing Britt. **pont-* (as in PONTYPWL SO2800 and many others) and Britt. **kagro-* (W *caer* "fort", as possibly in CHERHILL in Wiltshire) which would yield the modern form.

ROTHBURY ['rɔtbarɪ] (this pronunciation is probably dying out in favour of ['rɔθbərɪ]) NU0501 (*Routhebiria* ca 1100, *Rodbery* 1204; *Roburiam* 1210-2 represent a selection of the old forms) has been one of the standard-bearers for Scandinavian settlement in Northumberland because of the comparison with ON *rauðr* "red" as the first element of the place-name, also attested as a personal name as in RAUCEBY TF0245 in Lincolnshire. However, even if correct, one personal name does not make a Scandinavian settlement pattern, and the second seems uncontroversially to be OE *burg* "fort". Other authorities suggest an English personal name for the first element, such as *Hrōþa*. It may be that *Roth-* reflects Britt. **ratis* "fort" (as in ROATH in Cardiff, and ROTHMAISE in Aberdeenshire); if so, we could have another pleonastic formation. To the north of Rothbury is ADDYCOMBE NU0502 "Addi's ridge" (one *Addi* is recorded as having associations with the Bishop of Hexham in the late 7th and early 8th century). DEBDON LAKE NU0602 to the north is a creation of Lord Armstrong, who dammed the DEBDON BURN, a suitable location as it is a deep valley (OE *dēop* "deep"). NELLY'S MOSS LAKES NU0802 (and adjacent squares) are also new features, but the name may refer to an earlier area of boggy terrain (OE *mos* "bog, fen"). There is also a farm DEBDON NU0604, but the etymology suggests that the name of the valley came first. DEBDON PIT COTTAGE NU0704 and DEBDON WHITEFIELD NU0804 are secondary, the former named after the old coal workings and the latter after WHITEFIELD EDGE NU0803. Some names have disappeared from the maps altogether, such as MOUNT HEALEY NU0603, a different place from nearby HEALY NU0900 but probably with the same etymology (maybe an original dative used as a locative: OE *hēa(n)* + OE *lēage* "at the high clearing"). There was also a PAUPERHAUGH near here NU0603, and it is not clear to me which place the early forms refer to (see below). BIELDY PIKE NU0604 (probably a hill with a cairn and shelters for sheep, Nthb *beeld* + *-y*) and SWALLOW KNOWE NU0705 (one of the earliest recorded birds in English, OE *swealwe*) are to the east of the Debdon Burn. Another hill named after a bird is nearby GOWK HILL NU0904 (ME *goke* "cuckoo"). Other hills are named after animals: WOLF HOLE NU0903 (WOLF HOLE LETCH is the small stream which flows to the south-west), LAMB CRAGS NU1003 and LAMB HILL NU1004, none particularly troubling from an onomastic point of view. SHIRLAW PIKE NU1003 is a hill with an old cairn (typical of hills called *Pike* in Northumberland) which was either "bright or brilliant" (OE *scīr*, adjective) or a place for the shire moot (OE *scīr* æmoot, meeting place").

On the other side of the river, NEWTOWN NU0300 seems to be a transparent and relatively recent place-name, but it dates back at least to the 13th century (*Newtown* 1248, *Le Neuton* 1309). CARTERSIDE NU0400 could be the hillside of the *Crettinga* "Cretta's people", but this is purely by comparison with Cartington (see above). WHITTON NU0501 (*Witton* 1228) is probably "white farm" (OE *hwīt*), the motivation being either the character of the land or a whitewashed building. This has given rise to a number of secondary names, such as WHITTON GLEBE NU0500, WHITTONDEAN NU0500, WHITTON HILLHEAD NZ0499. There is a weir on the river near Whitton which provided the head of water for THRUM MILL NU0601 on the northern bank. The word is certainly a local word for a drumming noise, which could well be the motivation, but it is also the thread end of a weaver's warp and thrums or waste ends of threads were sold as bag ties, so one might imagine that mills were more often named after what they produced. On the southern bank, WAGTAIL FARM NU0700 seems to have a perfectly transparent name (*wagtail* is recorded in English since the 16th century). There is an Iron-Age camp just downstream and cup-and-ring marks to the south-west by the WAGTAIL BURN (the name derived from the farm or vice-versa?), but there does not appear to be any connection. To the high ground to the south, GARLEIGH MOOR NZ0699, GARLEIGH HILL NZ0699 and GARLEIGH CRAGS NZ0698 look as though they contain OE *gāra* "gore, triangular strip of land", as well as the word for clearing (OE *lēah*). Indeed, the piece of land enclosed by the road and the path past LORDENSHAW NZ0598 "lower valley copse" (ME *lagher*) does have a rough triangular shape. (Incidentally, if the etymology of *Lordenshaw* is correct, it cannot be earlier than ME, as *low* and *lower* are borrowings from ON and only became current after the OE period.)

BRINKBURN NZ1198 (*Brincewelæ* 1104-8, *Brinkeburne* ca 1120) with its priory must be among the most significant settlements along this stretch of the Coquet. It is on a bend in the river which provides natural defences and, as there are the remains of an Iron-Age fort which originally protected the promontory, it is safe to assume that the priory took over an earlier settlement. The name is usually taken to be Brynca's burn (Brynca is recorded in *Liber Vitae Ecclesiae Dunelmensis* "The Book of the Life of the Church of Durham"), and this person would then have given his name to BRINKHEUGH NZ1298 as well. BRINKBURN HIGH HOUSE NZ1199, BRINKBURN LODGE NZ1199 and BRINKBURN MILL NZ1198 would be secondary names. BRINKHEUGH NZ1298, MIDDLE HEUGH NZ1198, THORNEYHAUGH NZ1098, LONGHAUGH NZ1099, GLEADHEUGH WOOD NZ0999 (first element probably OE *gleoda* "kite"), and THISTLEYHAUGH NZ1398 are all names on the southern bank of the Coquet (OE *halh*).

WELDON NZ1398 (*Welden* ca 1250) is where the A697 crosses the Coquet, hence WELDON BRIDGE NZ1398 (the old bridge was built around 1760). There are several valleys with streams in the immediate neighbourhood which might have provided motivation. To the immediate south is the watercourse which flows through LINDEN GILL (OE *lind* "lime-tree" plus OE *denu* "valley") with GHYLLHEUGH NZ1397 on the eastern side (one imagines OE *halh* "land by a river" rather than OE *hōh* "height" as the landscape is relatively flat). The spelling *ghyll* appears to have been introduced by Wordsworth and used to spell names in the Lake District, and the word is ME *gille*, derived ultimately from ON *gil* "deep glen". One branch of this nameless watercourse rises near TODBURN EAST NZ1295, with TODBURN WEST NZ1195 over towards the TOD BURN (clearly the source of these names, reflecting ME *tod* "fox"), another candidate for the *wel*- element of WELDON NZ1398. Another interesting feature of this stream is that it flows through WHOLME GILL NZ1194, providing another "gill"-name. However, although WHOLME NZ1094 could reflect ON *hváll* "round hill" as in WHALE NY5221 in Cumbria, it is also possible that it continues OE **hwæl* "whale" (as in WHALLEY SD7336, *Hwælleage* 798, in Lancashire) and the OE cognate would account for the form better, the *-m* possibly reflecting the dative plural used as a locative. The stream also flows past GARRETT LEE WOOD NZ1196, which takes its name from GARRETT LEE NZ1096, a farm just to the east. Originally, it was "Gerard's clearing" (*Gerardesley* 1296), but now most of the woodland has been cleared.

CHAPTER SIX
6. An excursion around Kidland

The hills to the north and east of the Coquet provide excellent walking terrain, and there are a number of published suggestions for routes. One source accessible to everybody is (http://www.cheviotwalks.co.uk) which describes, among others, a circular walk around KIDLAND FOREST, a large area of woodland which seems to have belonged to Newminster Abbey in Morpeth in the medieval period. The name may be earlier and is usually taken to be "Cydda's land" (on the basis of forms such as *Kideland* 1271); there is one Cydda recorded in PASE, a Mercian who would likely have had more to do with KIDLAND in Devon, but at least the name is possible. Our route (we do not follow the walk exactly) begins along CLENNELL STREET, another drove road across the border. CLENNELL HALL NT9207 down by the banks of the River Alwin may be on the site of the old village (*Clenhill* 1242), one of the "Ten Towns of Coquetdale" in the Middle Ages, but the name means "open or clear hill" (OE *clǣne*), so perhaps it referred to one of the settlements, maybe the remains on CLENNELL HILL NT9308 (a classic pleonastic formation). Near Clennell Street itself are CASTLE HILLS NY9102/9207, the site of an Iron-Age fort, and there are many more old settlements in easy reach.

It is possible to cross the River Alwin and climb THE DODD NT9209 (see the discussion of this name above), with ROOKLAND HILL NT9308 and OLD ROOKLAND to your right. This probably contains the word for "rook", OE *hrōc*, which may also have been used as a personal name, so either "land where rooks are" or "land belonging to Rook". ROOKLAND NT9407 is a relatively modern farmhouse on the other side of SILVERTON HILL NT9308 (*silver* is usually taken to refer to the colour of the grass so "silver hill hill", another pleonastic formation if we assume *-ton* for original *–don* "hill"). LOUNDON HILL NT9408 also contains this element, likely reflecting Britt. **dūn(om)* in view of its Romano-British settlement, but the first element is more problematical. *Lound-* usually reflects ON *lundr*, as in LUND in Sweden, and this would have the meaning "grove, copse" which is not impossible, but unlikely in view of the location. Some may see a parallel in the name of LONDON TQ3079, but this is also not fully explained. On the way up THE DODD, a path veers off overlooking the PUNCHERTON BURN towards PUNCHERTON NT9309, where the existence of old forms (e.g. *Pun(t)chardon* ca 1250) suggests there was indeed a settlement here in the Middle Ages named after a Norman family, often mentioned in early Northumberland records, from Pontchardon in Normandy. PUNCHERTON HILL NT9209 is to the north-west and probably contains the family name rather than one derived from OE *dūn* "hill" or the like. GILLS LAW NT9409 is just to the east, and does contain a word for hill (OE *hlāw*) as well as what looks like a genitive and therefore probably a personal name (it may be the same *Gille* as in GILSLAND in Cumbria NY6366 or East Lothian NT5484).

Below THE DODD to the west is KIDLANDLEE NT9109 ("clearing" is still appropriate as it is in the middle of the forest) with KIDLANDLEE DEAN being the valley to the south. The ALLERHOPE BURN has early forms (e.g. *Alrehopeburn* 1240) which point to OE *alor* "alder" as the first element. It rises on WHOLHOPE HILL NT9311 (WETHER CAIRN is the name by the triangulation pillar NT9411), not to be confused with WHOLEHOPE NT9009, but as they are so close together, and given the difficulties in identifying names in charters with physical locations anyway, who is to say that they have not been mixed up already? There are early forms, including *Holehope* 1233, *Hollop* 1296, which point to OE *hole-hop* "hollow hope", and which could be applied to either place. As for as the change in spelling from initial *h*- to initial *wh*- is concerned, OE *hāl* and ME *hale* (as in *hale and hearty*) underlie MnE *whole*, so that should not concern us unduly. The medieval estate seems to have been divided up into "hopes". So as we progress through the forest, we encounter MILKHOPE NT9211 (OE / Anglian and ME *milc*) for dairying, and DRYHOPE NT9211 (OE *drȳge* "dry" is often used to refer to higher, drier ground), as well as NETTLEHOPE HILL NT8911, which may well contain OE *netele* "nettle", a plant with a variety of uses in the Middle Ages, but could just possibly be OE *nēat* "cattle" plus OE *hyll* "hill", which would link up thematically with HOG LAIRS NT9113 (ME *hogg(e)* "domestic pig / wild boar / young sheep" + ME *lare* "soft ground") and WETHER HILL NT9012 (OE, ME *weðer* "castrated male sheep"). The superficial link with FLESH SHANK NT8910 just to the south, however, is probably illusory: this is more likely to be ME *flashe* "marshy pool". There are two hills called INNER HILL NT9111 and NT9210 and one MID HILL NT8912 as well as a MIDDLE HILL NT8712 (which could always have been "mid hill hill") on the periphery of the presently forested area. I am unable to suggest what might be going on here, and SNEER HILL NT9011 is also a mystery, unless it is simply ME *sneare* "snare (for catching small animals)". HEIGH NT9111 [haið] suggests OE *hǣþ* "heath" or *hīehþo* "height" on the basis of the pronunciation (which usually counts for something) rather than comparison with HAIGH SD6009 in Lancashire, which relies on the modern spelling (which often counts for little). It is also located on what could have been a high heath between the YOKE BURN (perhaps OE *geoc* "yoke" on account of its shape) and the LINDHOPE BURN (if the *-d-* is original, this is probably OE *lind* "lime-tree", otherwise OE *līn* "linen, flax"). On the Yoke Burn were thought to be the remains of MEMMER KIRK NT9212, a chapel built for the estate, but this has been proved by excavation to be a 14th-century farmstead and the location of the chapel is unknown (www.pastscape.org.uk) as is the etymology of the name.

It is probable that the estate extended north of the presently forested area and therefore included YARNSPATH LAW NT8813 (*Hernispeth* 1233, probably OE *earnes pæð* "eagle's path"), BLOODYBUSH EDGE NT9014 (there was a battle here in 1585 which may have provided the motivation for the name), and USWAYFORD NT8814 (*Useyfoord* 1743) on the USWAY BURN (*Osweiburne* has no date, but appears in a Newminster document, so is medieval), which flows from DAVIDSON'S LINN NT8815 (probably a personal name with OE *hlynn* "torrent") to join the Coquet at Shilmoor. The name *Ōsa* is a shortened form of the many OE names beginning in *Ōs*- (there is even one Osa recorded in PASE). All the names derived from it are pronounced ['uːzɪ], like the machine-gun. DAVIDSON'S LINN is on the SALTER'S

Road, a drove road named after the route from the salt pans on the coast to Clennell Street and the border. It follows the upper reaches of the River Breamish, of which more below. Cushat Law NT9213 (*Cousthotelau* ca 1200) dominates the landscape to the south and it is hard to imagine wood pigeons on the hill today, but that it what the name seems to mean (OE *cūscote* "wood pigeon"). There are several boundary stones between here and Sting Head NT9312, which is explicable as the head of the Sting Burn, which could have been named after the Sting Cross which Dixon reports as having been located on a ridge between Cushat Law and Hogdon Law NT9412 (presumably "hog hill hill"), where posts or stumps (OE *stæng*, Nthb. *sting*) were maybe used as boundary markers. The name Saughy Hill NT8910, reflecting Nthb. *saughy* "abounding in willows" (the same word as in the famous Glasgow thoroughfare Sauchiehall Street, and OE *sealh* "willow"), suggests, along with the wood pigeons, that the area used to be forested much more widely.

If the return route is along the Usway Burn, we pass Batailshiel Haugh NT8810, which is almost unique in having solid evidence for a personal name — the Newminster Cartulary (ca 1225) describes it as *logia quondam Wilhelmi Bataile* "a house at one time of William de Bataille". The Usway Burn flows through a deep valley towards Shilmoor between another Inner Hill NT8708 and Copper Snout NT8809. Although, like Welsh *trwyn* and Gaelic *srón*, *snout* is occasionally used for a projecting hill, there does not appear to have been any copper mining hereabouts and this may be a popular etymology, perhaps of OE *copp* "summit". Those not wishing to walk back along the road may join the Pass Peth, a path through the pass, and walk over the southern slope of The Knocks NT8907, which like Knocklaw NU0601 and Knockshield NY8350 may possibly reflect Britt. **kunuko-* "small hill" (Welsh *cnwc*). You cross over the watercourse Passpeth Sike and look up to Middle Moor NT9007 and Green Side NT9007, the hillside on your left, before crossing the curiously spelled Pottstle Sike (I have no intelligent suggestions for the first element, unless it is *pott still* "a still where heat is applied directly", recorded from the 18th century onwards). This is, after all, a part of the country where illicit distilling was widespread, and the remains of Black Rory's famous still can still be seen below Davidson's Linn NT88331517. Fortified by this thought, we can then rejoin the road above Midge Holes NT9006 (this seems like a modern name, but the existence of OE *mycg* "midge" makes early formation possible).

CHAPTER SEVEN
7. The Cheviot from the East

From the right vantage point, the THE CHEVIOT NT9020 ['tʃɪv(j)ət] (the pronunciation ['tʃiːvjət] with [iː] is from the spelling, but now almost universal) may be glimpsed from as far south as Newcastle and the hills above the Tyne valley. The earliest names include *Chiuiet* 1182, *Chyvietismores* 1244, *Chyviot* 1250, *Chivyet* 1251, and it is generally agreed that the name is probably pre-English and obscure. Similar names appear in the West Riding: CHEVET is an element in a cluster of place-names SE3415 (and adjacent squares), which are on a low hill south of Wakefield (there is even a triangulation point) with early forms such as *Cevet* DB, *Chivet* 1153-5; THE CHEVIN SE2044 (*Scefinc* ca 972), a hill between Otley and Guiseley also gives its name to a cluster of locations in the vicinity, and is taken by Ekwall to be the equivalent of Welsh *is cefn* "below the ridge". This assumes the well-known change from [k] to [tʃ] before front vowels called the English Palatalization. Apart from being well attested in English words (such as *cheese*, OE *cese* borrowed from Latin *cāseus*) the change is also seen in English place-names borrowed from British (so **kaito-* "wood" often becomes *Chet-*). CHEVENING TQ4857 in Kent is also on the southern slope of a ridge. So the Welsh word *cefn* "back, ridge" (OC *chein* glossing Latin *dorsum*, MB *queyn*) seems like a good place to start to explain the name of THE CHEVIOT. *Cefn* appears in a host of Welsh place-names indicating ridges of all sizes from CEFN CRIBWR SS8582 to CEFN CENARTH SN9676, and CEFN YR YSTRAD SO0813, and probably also in the Gaulish name of the CEVENNES in France (Caesar's *Mons Cevenna*, Pliny's *Cebenna*). Taking the place-names and the words in Welsh, Cornish and Breton together, the methods of comparative philology allow us to reconstruct a British form **kebno-* "back, ridge". If we further compare SHEVIOCK SX3754 in Cornwall, also on the southern slopes of a ridge, which is reckoned to be identical with YSCEIFIOG SJ1571 in Flintshire, and YSGEIFIOG SH4773 on Anglesey, and the place in the Book of Llandâf noted by Ekwall called *Tref ir isceiuiauc* which is glossed as *villa proclivii* "the house of the slope", we can reconstruct a **keb-yāk-a* "sloping" with the same root **keb-* but a different extension. It was normal for early Indo-European languages to have a series of root extensions which sometimes became fossilized in the modern languages (so Latin *flōs* "flower" with an <s> corresponds to Irish *bláth* "flower" with a <t> and English *bloom* "flower" with an <m>) and we might reasonably assume that the root **keb-* in Brittonic had different extensions too, perhaps **-yāt-*, which would yield Cheviot quite comfortably. Alternatively, the root extension may have been changed by analogy with the (river) TEVIOT just over the border (here the root seems to be **tam-* as in the River THAMES, earlier *Tamesis* recorded in Caesar's account of 51 B.C.). There is also the idea that the second element might be OE *geat* "gate, gap in the hills", but this does not take the other forms discussed above into account, and ignores the fact that we are dealing with a hill, not a gap. If there is a connection with "gaps" it may be via the name of CHEW GREEN NT7808, as the element Chew often appears to mean "deep narrow valley", exactly as the Coquet valley is at this point. Finally, there is a suggestion I cannot dismiss entirely which appears in Tomlinson (1888:480), that Cheviot is a compound of *cefn* "ridge" and *ôd* "snow". This fits in well with the topography — the Cheviot is often visible capped with snow — and the Welsh word does exist (there is also a verb *odi* "to snow"), alas without a convincing etymology, but it is a tempting explanation.

All this will provide plenty of food for thought as you are tramping through the countryside. But how to get there? The gateway to the Cheviot is WOOLER NT9928 (*Wullovre* 1186, *Welloure* 1196, 1203, *Wllovera* 1199, etc.). There seems to be very little archaeological evidence for the OE period associated with Wooler, although this is almost certainly when the name was formed. It could mean "Wolf's / wolf's bank", probably reflecting a personal name *Wulf* rather than OE *wulf* "wolf", although one never knows, and there is a preferable alternative. The second element appears to be OE *ofer* "flat-topped ridge" and we should note that the earlier settlements were on the high ground to the west rather than down by the river. The hill nearest to Wooler is called HORSDON HILL on the maps, and on its southern slopes is a wishing well or spring (*The Wishing or Pin Well* in Tomlinson's account, p. 476). OE *wella* "spring" could easily become **wolla* / **wulla* because of the initial [w], and if this is the origin of the first element the name would mean "flat-topped ridge with a spring". On the eastern slopes of Horsdon Hill is the tract of land called THE KETTLES NT9827 which "abounds in pot-like cavities" (Heslop) called *kettles* after their shape. The name of HORSDON HILL NT9827 (*Horsdon* 1865 OS map) appears to be identical with the one Tomlinson (p. 476) calls *Horsdean* "on which the statute fairs for sheep and cattle are annually held", which might explain the first element (OE *hors* "horse"), but the fact that we are dealing with a hill suggests OE *dūn* "hill" rather than OE *denu* "valley". The camp or hill-fort was a large one and would justify its alternative name of MAIDEN CASTLE (Britt. **maglo-dūnom* "great fort" + more modern *castle*), not an uncommon way of referring to British hill-forts (e.g. MAIDEN CASTLE in Dorset), but it could also be secondary, which is probably the case with MAIDEN CASTLE in Cumbria, a Roman fortlet. The other name given to the location is GREENSIDE, i.e. "green hillside", and I surmise that this belongs to a later stratum of nomenclature.

The end of a path now known as St. Cuthbert's Way (the route on which his body was carried before final interment in Durham Cathedral) skirts the hill to the north of the fort and passes WAUD HOUSE NT9827, which Tomlinson (1888:476) calls "the Wadhouse" and which, as it is not marked on the 1865 OS map, we must assume is late 19th-century. The seat-like rock called the KING'S CHAIR is to the south-east. This name is not unique. It is parallelled in KING'S CHAIR HILL NT9035, near to FLODDEN HILL NT9135, on which there are earthworks which look suspiciously like an Iron-Age fort. There is also a spring near both locations: Sybil's Well on Flodden Hill and the Wishing Well or Pin Well already mentioned on the slopes south of Kettles Fort. Further down the slope, the road leads to EARLE [jɛrl] NT9826 (*Yherdhill* 1242) a "hill with an enclosure" (OE *geard*), which also gave its name to various locations. We note EARLE HILL NT9825, and EARLEHILLHEAD NT9726 (both pleonastic formations). Then there is EARLE WHIN NT9826 (probably named after the vegetation *whin* "gorse", compare WHIN FELL NY5643 and NY1325, both in Cumbria, but it is unusual as a second element). Finally, there is EARLE MILL NT9926 (the mill is 18th-19th century, so the name may be modern as well) down by WOOLER WATER. KENTERDALE HILL NT9727 is unusual in containing the element *-dale*, usually associated with areas settled by Scandinavians, but it is probably OE *dāl* "valley" in this case. Presumably, the valley of what is now the Humbleton Burn is meant, so the first element *Kenter-* could contain the original name of the watercourse, paralleled in the name of the rivers KENNET, KENNETT and KENT, all taken to represent Britt. **kunētio-* which we see in the name of the Roman town at MILDENHALL SU2169 in Wiltshire, *Cunetione*. The dogs implied by the name (Welsh *ci*, plural *cwn*) are likely to have been river dogs, i.e. otters. The genitive in nearby BROWN'S LAW NT9727, which overlooks the HUMBLETON BURN, suggests a personal name. The burn bifurcates just west of BELL'S VALLEY NT9527, and the upper reaches are marked THE TROWS NT9527, probably after trough-like depressions in the stream bed, although the existence of southern Scottish place-names such as TROSTRIE NX6557 reflecting Britt. **trāns* "cross, across" (= W *traws*, etc.) raises the possibility that the same element is reflected here. FREDDEN HILL NT9526 (I have no suggestions for the first element, the second is probably OE *dūn* "hill") and (yet another) WATCH HILL NT9526 are to the south.

WOOLER WATER, originally called CALDGATE BURN according to Tomlinson (hence COLDGATE MILL NT9925), is actually a continuation of the HARTHOPE BURN (*Herthop* 1305) and the Harthope valley (probably "blind valley where stags are found" with OE *heorot* "stag"). This is a good place to start exploration of the Cheviot itself. Just below Earle is WALKERWALLS NT9825, a row of cottages now converted into one building, which probably refers to a former owner, maybe even the Thomas Walker mentioned on the Grade II listed gravestone in nearby Ilderton churchyard. These place-names usually have a genitive as first element, however. We pass MIDDLETON HALL NT9825, one of the three Middletons (*Tres Midiltonas* 1289), which also included NORTH MIDDLETON NT9924 and SOUTH MIDDLETON NU0023. Confusingly, MIDDLETON OLD TOWN NT9924 may have been the one in the middle (*Midilest Midilton* 1296), and the site of another medieval village was probably to the south of the road near Middleton Hall. They all mean "middle farm or enclosure". A valley called MIDDLETON DEAN NT9922, is dominated by MIDDLETON CRAGS NT9721 above. We now climb over the shoulder of yet another WATCH HILL NT9825 (see the discussion of GREAT WATCH HILL NY7075) with the park of MIDDLETON HALL and GRIMPING HAUGH NT9825 below. *Haugh* is clear enough and the slope is steep enough to make one think of the rare verb *grimp* (from French *grimper* "to climb") as a possibility; alternatively, there is the sparsely attested *grimpen* (as in *Great Grimpen Mire*), so further investigation is needed. The house, originally a shepherd's cottage, on the ridge above the road is called

SKIRL NAKED NT9725. A similar phrase is listed in Heslop's dictionary, namely *skirl-neak't* "stark naked", of which he says "a common expression in Coquetdale; usually applied to children", so perhaps this was a derogatory term for the previous inhabitants of the cottage. KINGSTONE DEAN and KINGSTONE PLANTATION NT9725 are just behind, seemingly modern coinings, but perhaps there is a connection with KING'S CHAIR (see above). On the other side of the dip, SWITCHER seems to derive its name from SWITCHER WOOD NT9725, a source of the switches once used, for example, as riding whips. The road then crosses a bridge over the CAREY BURN (*Care Burn* in Tomlinson p. 478) where it joins the Harthope Burn, and this name in contrast to the others does appear to be old (see the discussion of CARRYCOATS HALL NY9279 above). The stream is fed by the waters of the COMMON BURN (this name must be ME or later, as ME *co(m)mun* replaced OE *gemǣne*) and BROADSTRUTHERS BURN (ME *strother(s)* "marsh(es)") just north of LUCKENARKS NT9525. If it is to be taken at face value, this could mean "locked / enclosed bins" (ME *ark* was used to refer to "a large box for storing meal or grain", the first element being the old past participle of ME *louke* "to lock, enclose"). BROADSTRUTHER itself is in NT9424, to the west of HAZELLY BURN (the frequency of adjectives in *-y* in Northumberland place-names has been noted above, compare also nearby STEELY CRAG NT9524 with OE *stigol* "ridge" + *-y*). CARLING CRAGS NT9524 overlook HOLLY CLEUGH NT9524. *Carling* may be a personal name, or derive from a superstition regarding witches (ME *carling* "old woman, witch" occasionally occurs in place-names), or be associated in some way with Carling Sunday (the fifth Sunday in Lent, when it was customary to eat parched peas or *carlings*).

Back on the road up the Harthope valley, we pass CORONATION WOOD, which appears on the map between 1980 and 1994, so is very modern). SNEAR HILL NT9624 (probably ME *sneare* "snare") rises behind. BRANDS HILL NT9723 with its myriad of Roman-period native settlements is on our left (*Brand* may have been a later Anglo-Saxon settler or the hill may have been named after the legendary ancestor of Anglo-Saxon kings, the son of *Bældæg*). LANGLEE NT9623 "long clearing" is further up the valley with LANGLEE CRAGS NT9622 to the south. EASTER DEAN and WESTER DEAN run down either side of the crag (the names sound Scandinavian, but they could equally be OE **wester* or *westerra* "westerly" as in WESTERTON SU8807 in Sussex and *ēasterra* as in EASTERTON SU0255 in Wiltshire, and the *dean* element suggests non-Scandinavian origin anyway). The LEECH BURN (OE *lǣce* "blood-sucking worm") runs parallel to the Harthope Burn before joining it just above Langlee, and separated from it by a piece of higher ground called THE SHANK NT9622 (OE *sc(e)anca*: we have seen how common it is to use words for parts of the body for natural features). On the other side of the Harthope Burn, a similar ridge separates the PINKIE SIKE from the main watercourse and is marked PINKIE SHANK NT9523 (probably Nthb. *pinkey* "small" rather than another body-part, and applied to the stream first, as this ridge seems larger).

For those who wish to climb to the summit of the Cheviot, a good route begins where the HAWSEN BURN and HARTHOPE BURN converge NT9522, just past COCKSHAW SIKE (OE *cocc* "game-bird" but there is no trace of the copse now). The HAWSEN BURN is *Hawsden Burn* in Tomlinson, and this makes more sense, reflecting either OE *haga* "enclosure" or OE *haga* "haw", or just possibly OE *hafoc* "hawk", the second element being OE *denu* "valley". Yet another COLD LAW NT9523 "cold hill" is on one side and BLACKSEAT HILL NT9422 "black hillside hill" is on the other. A tradition that there was "a Celtic town" on the lower slopes may justify OE *sǣte* "farm" as an explanation for the second element, otherwise it is probably ME *sīde* "hillside", the element OE *blæc* "black" being easily justifiable in view of the peaty soils. Names such as STICKY BOG NT9423 and RUSHY GAIRS NT9424 (Nthb. *gair* "grassy spot surrounded by bent or heather") give one a good idea of the terrain to be expected. If we take the path up to BROADHOPE HILL NT9323 "wide valley or piece of land hill", there should be a track leading almost due south to SCALD HILL NT9221. It is fine to imagine Scandinavian poets (ME *scald*) up here chanting verses, but I think it is more likely that manure was dropped on the land from carts in small heaps which were subsequently *scaled*, or "scattered evenly over the surface" (see Heslop). From here it is a straightforward, but steady climb up to the summit of Cheviot itself and the extension of the Pennine Way.

Further up the valley, LANGLEEFORD NT9421 really does have a passable ford, but the right of way is on the north side of the burn (LANGLEEFORD HOPE NT9320 is further upstream still). CAT LOUP NT9421 is the name of a narrow gorge, a "cat's leap" (Nthb. *lowp, loup* "a leap", see Heslop, who also mentions the *dog-loup* "a narrow passage between two adjoining but detached houses"). To the south, HOUSEY CRAGS NT9521 (the presence of a Roman period farm nearby, as well as the Iron-Age settlement to the south may have motivated ME *hūs / hous* + *-y*), and LONG CRAGS NT9521 (a rock outcrop easily elongated enough to justify the name) overlook the valley. TATHEY CRAGS NT9621 (ME *tathen* "to dung", cf. ON *tað* "dung, manure", cf. SCALD HILL above) are on the other side of BROAD MOSS NT9621, "wide bog, fen" near the site of a prehistoric settlement (all these habitations at such an elevation are indicative of a warmer climate at the time). HARTHOPE LINN NT9220 is in Heslop's words "a succession of pretty waterfalls overhung with native wild wood and fringed with ferns"; presumably this location was also frequented by deer. The burn rises just below SCOTSMAN'S KNOWE NT9018 with SCOTSMAN'S CAIRN NT9019 just to the north (both these are on the 1865 OS map, but have a modern ring to them).

The famous circuit would now take you south and east to COMB FELL NT9118 (doubtless OE *camb* "comb, ridge") and from there north-east to HEDGEHOPE HILL NT9419 (probably OE *ecg* "crest", the initial [h] being the result of dissimilation). The path back to the car takes you north-east again past KELPIE STRAND NT9520 (ME *strand* "rivulet" of obscure origin, so this name may have referred to one of the tributaries of the Threestone Burn originally). Kelp was burned on

A view of Hesleyside NY8183 from the public road, set in parkland and not on a "hazelly hill", even though that is what the name means.

The house at Mantle Hill NY8184 can just be seen behind the trees. If the name is another example of a pleonasm, or onomastic tautology, it would mean "hill hill hill". The public bridleway leads to Pundershaw NY7880 "the copse of the keepers of the pound".

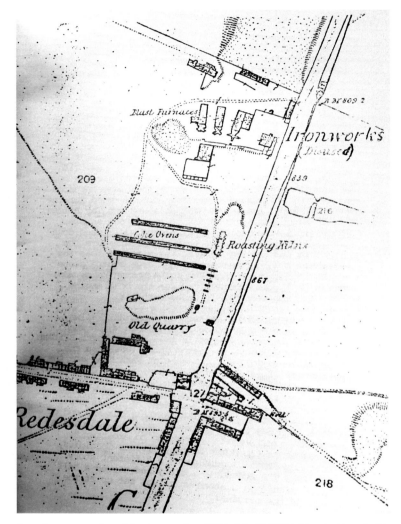

A contemporary plan of the iron works at Ridsdale NY9084, bisected by what is now the A68. The name is probably just a local pronunciation of Redesdale, as the legend in the bottom left-hand corner suggests.

Although commonly assumed to be a castle, this is in fact the former engine house for the Ridsdale iron works. It housed two beam engines which supplied three blast furnaces.
(© Peter Maddison and licensed for reuse under Creative Commons.)

Two views of Yardhope Man, near the course of the Roman road at Yardhope NT9201, possibly a depiction of the local British war-god Cocidius.

Northern English place-names in -by (ON býr "farm")

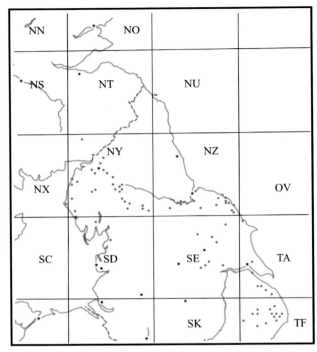

Note the absence of such names in Northumberland
(the equivalent would be OE -hām, -tūn, etc.).

The *pench-* element of Penchford NY9497 is a bit of a mystery. If it is the same as Penge TQ3470 in Middlesex, it could be another Brittonic survival and mean "end of the wood".

The position of The Raw NY9498 on a slope makes a derivation from Brittonic *rī–wā "slope, hillside" (Welsh *rhiw*) at least plausible, and invites a connection with Cumrew NY5450 in Cumbria, and similar names in Cornwall.

Two ramparts and a ditch, part of the Iron-Age hillfort on Castle Hill near Callaly NU0509 (written as *Calualea* 1161, from OE *calfa* "calves" + *lēah* "clearing"). A modern name such as Castle Hill is likely to indicate an old hillfort.

West Hills hillfort.

Great Tosson hillfort.

Lordenshaws hillfort.

The location of the modern town of Rothbury NU0501 (pictured on the previous page), would have been surrounded by hillforts in the Iron Age. Great Tosson hillfort NU0200 (also previous page) is located on the southern side of the Coquet. To the north of the river is West Hills NU0302 hillfort, and just to the east of that is Old Rothbury NU0401. Lordenshaws NZ0599, pictured above, is on the high ground south of the river. Perhaps this cluster of hillforts gives us a clue to the origin of the name of Rothbury — a pleonastic formation meaning "fort(s)fort".
(All the aerial views are © Google Earth.)

The Drake Stone NT9204 surveys the countryside near Harbottle just like a dragon, and Latin *draco* "dragon" is indeed the most likely origin of the name
(© johndal and licensed for reuse under Creative Commons).

The Cheviot NT9020 is visible from far and wide in Northumberland, but the origin of its name is less obvious. It is definitely pre-English, and the first element may be connected with the Welsh word *cefn* "back, ridge". From many angles, the hill certainly looks like the back of a sleeping giant lying on his side. The second element may be linked to the Welsh word *ôd* "snow", motivated perhaps by the frequent appearance, except in the summer months, of snow on the summit.

A view of the new house at Batailshiel Haugh NT8810, the name of which is almost unique in Northumberland in having solid evidence for its origin in a personal name — the Newminster Cartulary (ca 1225) describes it as *logia quondam Wilhelmi Bataile* "a house at one time of William de Bataille".
(© Andrew Curtis and licensed for reuse under Creative Commons.)

the coast and may have been exported inland to help fertilize upland farms.

CHAPTER EIGHT
8. The Cheviot from the North

There are two routes to the summit of the Cheviot from the north: one via Henhole, the other via the Bizzle (or, as Tomlinson has it, the Bazzle), both reached by the College valley. The starting point for most will be Wooler (see above), but this time the direction of travel should be north on the A697. HUMBLETON HILL NT9628 (e.g. *Hameldun* 1169) is to the left and contains the element OE *hamola* found in HUMBLE HILL ("bare or hornless / dodded hill", see above) plus OE *dūn* "hill" (which may well reflect Britt. **dūnom* "hill-fort" in view of the Humbleton Hill fort), plus pleonastic *hill*. The village of HUMBLETON NT9728 probably takes its name from the hill, and the early 19th-century farm HUMBLETON BUILDINGS NT9728 probably takes its name from the village. There are two Roman-period native settlements on COLDBERRY HILL NT9627: OE *byrig* is the dative case of *burg* "fort" and often gives place-names in *-berry / -bury* (e.g. BLACKBURY CAMP SY1982 in Devon), whereas OE *ceald* "cold" is sometimes inevitable in Northumberland. The course of St. Cuthbert's Way is just north of here and runs over GAINS LAW NT9528, which is usually taken to reflect a personal name as in GAINFORD NZ1716 (one early form is *Gainesford* 1207).

To the north of this is HAREHOPE HILL NT9528 with Bronze-Age and early Iron-Age settlements, a piece of land (OE *hop*) where hares (OE *hara*) could feel at home. To the south of the summit is MONDAY CLEUGH NT9528, a ravine which, given its position, may contain the same element as nearby MINTO NT5620 and MINDRUM NT8432, as well as MUNDIE NO1114 and MUNDAY NO0720 further north, i.e. Britt. **moniyo-* (Welsh *mynydd* "mountain"). From here, we are afforded an excellent view of BENDOR NT9629 with its stone associated with the Battle of Homildon Hill in 1402 (but the stone is actually a Bronze-Age standing stone). In the case of AKELD NT9529 (*Achelda* 1169), we are probably dealing with a slope covered in oak trees (i.e. OE *āc* "oak" + *helde* "slope", rather than ON *á* "river" + *kelda* "well, spring"). The farm clearly owned property on the other side of the River Glen: AKELD STEADS NT9630 (also *Akeld Stead*) reflecting OE *stede* "place, site of a building", perhaps AKELD LODGE NT9530, built in the early 19th century. It also gives its name to AKELD HILL NT9429 (or perhaps this pleonastic formation came first). The wonderfully named GLEADSCLEUGH NT9529 probably reflects OE *glida* "kite". The RIVER GLEN (*in fluuio Gleni*, ca 730) falls into the Till in a square (NT9732) almost unique in its dearth of place-names (OX CLOSE PLANTATION seems to be the only one). For the river, the most likely etymology is Brittonic **glanyo-* "pure, white, fair" (cf. Welsh *glan*, which also means "holy"). This would be reasonable in view of the clear water, and also the sacred significance attached to rivers in pre-Roman Britain. There are also rivers called GLEN in Lincolnshire and Leicestershire, the latter possibly deriving from Brittonic **glenno-* "valley", which provides a less likely explanation for the Northumberland GLEN.

The magnificent hill-fort of YEAVERING BELL NT9229 is the next vantage point over the Glen, with YEAVERING NT9330 and OLD YEAVERING NT9230 below. This (or a site just to the north) is Bede's *Ad Gefrin*, ca. 730 AD, usually derived from a compound of Britt. **gabro-* "goat" (Welsh *gafr* "goat") and Britt. **brunnyo-* "hill" (Welsh *bryn*, etc.), doubtless with the hill in mind. However, this need not be the second element, as **dubro-sentu-* "water-path" gives Welsh *dyffryn* and **epo-sentu-* "horse-path" gives the name of the MYNNYDD EPPYNT, so **gabro-sentu-* "goat path" could yield *Gefrin* (and *Gabrosentum* appears in the *Notitia Dignitatum* or "register of offices" dating from the late 4th or early 5th century). NEWTON TORS is the name for the high ground to the south, and they appear individually as EASTER TOR NT9128 and WESTER TOR NT9027, which reflects OE *torr*, but this has to be a loan from Brittonic, as it has no related words in any other Germanic language, yet appears frequently in place-names in Devon and Cornwall as well as Gloucestershire, Cumbria, and here. TORLEEHOUSE NT9128, near to another Roman-period settlement, clearly contains the same first element, and the second may be either OE *lēah* "clearing" or OE *hyll* "hill" (as in KIRKLEY NZ1576, perhaps DINLEY NY8877, however improbable this may seem).

NEWTON is such a common English place-name (OE *nēowa tūn* "new farm") that many were distinguished in some way. So it is with KIRKNEWTON NT9130 but the *kirk* element is not necessarily evidence for Scandinavian settlement, as OE *cirice* "church" was often "scandinavianized" later, and the early form does not contain the word at all (*Niweton in Glendala* 1336, i.e. "new farm in Glendale"). Besides, the Anglian form of OE *cirice* may not have shown the palatalization anyway. WESTNEWTON NT9030 and WESTNEWTON BRIDGE NT9030 are later and onomastically transparent. This is where the COLLEGE BURN (also called the BOWENT) joins with the BOWMONT WATER to form the GLEN. Ekwall takes BOWMONT and BOWENT to be the same, and observant readers will notice the *-went* ending characteristic of the group of British river names like the ALWIN, discussed above. The earliest recorded spelling is in the phrase *fluvium Bolbenda* (ca 1050), but the spelling should not put us off, as -*b*- could be used to indicate [w], and [l] before a consonant is often vocalized (as in the local pronunciation of *cold*). If **bol-* is indeed the first element, it could reflect the IE root **bhel-* / **bhol-* for which the meanings "white" and "swamp(y)" recur frequently, pointing us to a meaning "white river" or "swampy river" for the Bowent. As far as COLLEGE is concerned, this is usually taken to be a compound of OE *c(e)ald* "cold" and ME *leche* "swampy watercourse", providing a nice semantic parallel with the Brittonic etymology. At any event, the existence of two names for the river suggests that the two linguistic groups (Brittonic and Anglian speakers) co-existed in the area. Above Kirknewton, ST GREGORY'S HILL NT9129 (there is another hill of this name in North Lancashire, but I have not yet found an association of any St Gregory with either location) and WEST HILL NT9029 both have Iron-Age forts and settlements. Across the College Burn, THE BELL NT9028 (also known as HETHPOOL BELL) with yet another fort and enclosure contributes to the evidence for the word *bell(e)* as "hill". HETHPOOL LAKE NT8928 seems like a pleonastic formation (OE-ME *pol* "pool" and *lake* mean roughly the same), the first element possibly being OE-ME *hæð*, ME *heþ* "open, uncultivated ground", i.e MnE *heath*. The name appears in nearby HETHPOOLBELL WOOD NT8928 (as in THE BELL above) and both HETHPOOL LINN NT9028 (OE *hlynn* "torrent"), and the deserted medieval village as well as the present-day farm logically derive their name from the lake, as does HETHPOOL MILL NT8927 on the southern side of the College Burn. One problem with deriving *Heth-* from OE *hæð* "heath" is the existence of GREAT HETHA NT8827 and LITTLE HETHA NT8828, two nearby hills both with forts on their summits; as both the hill and the fort are larger in the case of GREAT HETHA, one cannot help thinking that *Hetha* might refer to either the hill or the fort. If the latter, an etymology readily suggests itself, namely Britt. **sedos* "seat, dwelling" (W *hedd* now means "peace, tranquility"), which, if followed by an adjective beginning with another *d-*, such as **dagos* "good", would result in **sedo-dago-* "good dwelling", the regular development of which could easily be *Hetha*. The HETHA BURN rises on the eastern slopes of MADAM LAW NT8626, the first element of which has the look of a popular etymology (as *madam* is a MnE loan from French), perhaps for the Brittonic word for "mother" (Welsh *mam*, occasionally used in hill-names, e.g. MAM TOR in Derbyshire).

On the other side of the College Burn is HARROWBOG NT8927 (*Harrowbog* 1866 OS map, *boag* 1820), a ruined post-medieval farmstead, for which a connection with OE *hearg* "heathen temple", perhaps via comparison with HARROW TQ1488 in Middlesex and other similar names, has been proposed. The next place up the valley is WHITEHALL NT8826, which has a name which seems self-evidently to refer to a "white hall", but we must bear in mind that the farm itself dates from the 19th century, so, if the name is old, it is unlikely to refer to these particular buildings. There is considerable evidence of earlier settlement hereabouts, right back to the Iron-Age, and *hall* may also be a continuation of OE *halh* "alluvial land". HARE LAW NT9026 (probably "hare hill") looks down from the eastern side and SINKSIDE HILL NT8826 and LOFT HILL NT8725 (not a unique name, see above) from the west. There was an Iron-Age fort on SINKSIDE HILL, so perhaps ME *synke* "a pool or pit formed in the ground for the receipt of waste water, sewage, etc." explains the first element, as even these early inhabitants would have needed somewhere to dispose of their waste. For the second and third elements, we have the familiar pleonastic pattern of ME *sīde* "hillside" + *hill*. LOFT HILL, separated from Sinkside by the WHITEHALL BURN, probably reflects OE *lyft, loft* "air, wind, sky", and may therefore contrast with the unsavoury sanitary arrangements on the hill just to the north. On the northern slopes of Sinkside is TROWUP BURN, and the farm TROWUPBURN NT8976, the early forms of which (e.g. *Trollop* 1352) suggest that this was the farm (*hop*) where trolley-carts or trolls for fishing were made (i.e. *troll*). The burn is fed by a comparatively short stream called WIDE OPEN, which rises at WIDEOPEN HEAD NT8626, whereas a much longer stream runs parallel to the border and the Pennine Way from its source on BLACK HAG /

CORBIE CRAIG NT8623 (early MnE *hagg* "peat-bog, overhanging mass of peat", ME *corby* "crow", ME *craig* "rocky outcrop") and SAUGHIESIDE HILL NT8624 (cf. SAUGHY HILL NT8910, reflecting Nthb. *saughy* "abounding in willows" + ME *sīde* "hillside" + *hill*) further south. One would expect WIDEOPEN to be just that, wide and open, but it is in fact a "steep, bracken-filled valley" as Andrew Curtis (www.geography.org.uk) put it, so the etymology is a puzzle. THE TONGUE NT8724 and BLACKHAGGS RIGG NT8724 (ME *rigg* "ridge") separate us from the College Burn.

Near the village hall, one spur of the road crosses SUTHERLAND BRIDGE NT8824 (which does not appear on the maps until the 1950s) and passes through SOUTHERNKNOWE NT8824, built in the mid-19th century, before arriving at COLDBURN NT8924, where the COLD BURN (a modern version of *College*, if the etymology of the latter is right) joins the LAMBDEN BURN underneath COLDBURN HILL NT9024. DUNSDALE CRAG is just to the north NT8923, and although the hill farm at DUNSDALE is of 19th century date, there is also an Iron-Age settlement nearby, which may give us a clue to the name (Britt. **dūn(om)* "fort", Welsh *dinas* "fort, refuge, city"). The second element is probably OE *dāl* "valley", and this most likely refers to the course of the Bizzle Burn. Like THE BIZZLE on West Kielder Moor NY6499, the stream flows through a narrow gorge called THE BIZZLE NT8922. There is of course the word *pizzle* "penis", but it is difficult to see the motivation here unless it refers to BIZZLE CRAGS NT8922 (but there is nothing obvious to suggest it) and that argumentation does not apply to the West Kielder Moor location. What they have in common is the gorge, which narrows at the top rather like a thimble. There is a Welsh word *byslle* "thimble" which (unfortunately for the etymology) is not recorded until the early modern period. On this meagre corroborative evidence, it is only just possible to reconstruct a Britt. **bisso-legā* "thimble", however. At the top of THE BIZZLE, we have MID HILL NT9021 and BELLYSIDE CRAG NT9021 (OE *bælig* "bag" — the plural gives us MnE *bellows* — or this might be a popular etymology of *bell* "hill") on the eastern side and BRAYDON CRAG NT8921 and WEST HILL NT8921 to the west, putting Mid Hill in context, as it were. BRAYDON BURN flows through BRAYDON CLEUGH NT8821 to join the COLLEGE BURN. The modern spelling suggests that the second element of *Braydon* is OE *dūn* "hill" and that BRAYDON CRAG is earlier, giving its name to the ravine and the stream. The first element could then be Britt. **brigā* "hill" (W *bre*, etc.) with the whole name being revealed as yet another pleonastic formation "hill hill". There would be a partial parallel in BREDON HILL in Worcestershire (although as this has a hill-fort, the second element here is probably older Britt. **dūnom* "a fort").

If we had carried on along the valley road, we would have reached the point where the Braydon Burn and the College Burn converge, via FLEEHOPE NT8823 (perhaps OE *flēoge* "fly", one of the few words for which an Anglian form is recorded, namely *flēge*) at the mouth of the FLEEHOPE BURN, and MOUNTHOOLY NT8822, of which there are at least five in Northumberland. They are all taken to be from OE *munt* "mountain, hill" (reckoned to be from Latin *mons, montis* "mountain", but it could equally well derive from Britt. *moniyo- "hill") plus OE *holegn* "holly", also found as a name in its own right (HULNE NU1615). THE SCHIL NT8622 is not far away to the west, probably reflecting OE *scylf, scelf* "peak, crag" as in SHELL SO9559 in Worcestershire (although the border SCHIL is much larger). SMEDDUM SIKE NT8721 supplies the upper reaches of the College Burn from the west. As an explanation I note early MnE *smitham / smeddum* which signifies fine or small particles of substances such as coal, lead ore or shale, so perhaps this watercourse (OE *sīc*) saw some mining activity at one time. Those who wish to reach the summit of Cheviot by this route will turn east via HEN HOLE NT8820 (perhaps "wild fowl hollow", OE *henna* + OE *hol*), for which there are, unfortunately for us, no reliable old forms.

CHAPTER NINE
9. The Breamish Valley

The name of the River BREAMISH, which rises in the Cheviots and forms the upper part of the Till above CHATTON NU0528, sounds English, but the early forms, including *Bromic* (c. 1050), *Bremyz* (1293), *Bremish* (1532), then *Bromish* (1586), show a well-established alternation between -*o*- and -*e*- which nobody has been able to explain. If the -*o*- is not just a scribal error (always possible, as the scribes were often not familiar with the actual places they were writing about) and is original, one might think of OE *brōm* "bramble" with a sound change of [o:] to [i:] which Ekwall notes "in Scottish dialects" so that *broom* is pronounced as [bri:m]. The name would then be English, but that would leave the suffix to be explained. If the -*e*- is original, one could explain the name as Britt. **brēm-* (found in the name of the River BREFI, a tributary of the TEIFI, and maybe also in the old name of HIGH ROCHESTER NY8398, *Bremenium*) with the suffix **iskā* "water" (found in the name of the Rivers EXE, AXE, ESK and USK, etc.). The second solution is plausible, and a British name would fit in with the first element of Chatton (see below), but it is impossible to be sure.

BREAMISH HEAD NT9018 is on the southern slopes of SCOTSMAN'S KNOWE NT9018, which seems so straightforward that one immediately thinks that a popular etymology must underlie the name, but I have no suggestions. Nearby COLDWELL STRAND NT9018 "cold stream stream" (OE *wella* "stream" + ME *strand* "rivulet") and SHEDDING SIKE NT8918 (ME *shēding(e)* "parting, divison", perhaps of sheep) feed the Usway Burn and eventually the Coquet. The HARPATH SIKE looks like OE *herepæþ* "main road", with *here-* to *har-* as in Harbottle, so perhaps (along with the place-name evidence of Shedding Sike, where livestock may have been sorted) the course of a drove-road ran nearby. It joins the Breamish a little downstream NT9018, as do a couple of unnamed streams from the slopes below COLDLAW CAIRN NT9118 (i.e. "cold hill"). From the west, OUTER QUICKENING CLEUGH NT9017 and INNER QUICKENING CLEUGH NT9016 (probably containing *quickening* "couch grass", or a word for the *quicken-tree* "rowan") drain the slopes of BROAD SHIN NT9016 (OE *scinu*, but the sense "sharp slope of a hill" is 19th-century) to the north of the Salter's Road path (see above). WHITEWELL STRAND (as in COLDWELL STRAND above) flows down from SHIELCLEUGH EDGE NT9217, but SHIEL CLEUGH itself, a ravine where there must have been a hut, is further downstream. HIGH CANTLE NT9216 is a hill, so it is easy to suppose (if it is not simply ME *cantel* "corner, nook, rim" from medieval Latin *cantellus* "little corner", or OE *cantel* "buttress, support") on the basis of the many hills with a similar ending, that -*le* is derived from OE *hyll* "hill". The first element could then be *cant* (ME *cant* "edge, border", one source talks of *the cant of a hill*), or, if older, we might think of Britt. **kando-* "white, grey" (W *can* "white, brilliant", OB *cant* "grey", MB *cann* "full moon"), or Britt. **kanto-* "circle" (W *cant*, B *kant*, MIr *cét*). High Bleakhope in the valley NT9215 looks like a relatively modern building, but the name may be older; we also have LOW BLEAKHOPE NT9315 (below LOW CANTLE NT9315). The *hope* element could mean "small enclosed valley" here, as both the farms are near such features, or simply "piece of land". The first element is also open to interpretation: it is usually taken to mean "dark" (i.e. OE *blæc*), but there is also a northern word *bleak* "to make white or pale by exposure to light" and perhaps this was simply a farm where bleaching was done.

Between AINSEY KNOWE NT9115 (just possibly from OE *ānstig* "single track") and NAGSHEAD KNOWE NT9116, the AINSEY BURN is fed by the REDSCAR SIKE (perhaps a reference to the pinkish andesite which is occasionally exposed) and the FOULSTEP SIKE (OE *fūl* "foul, putrid", but why step?) before being joined by the RUSHY SIKE (ME *rushy* "with rushes") to fall into the Breamish. Downstream from this are BLACK CLEUGH NT9215 and CAT CLEUGH NT9215, ravines with streams draining the eastern slopes of Ainsey Knowe. HARESHAW CLEUGH joins the Breamish where it takes a turn to the north-east towards the place where the MARCH SIKE (ME *mersh, marche* "border") and the BROOMY CLEUGH NT9316 (early MnE *broomy*) feed the river. I cannot make anything of the first element of nearby SNUFFIES SCAR NT9415. The second element is a word apparently from ON *sker* "low reef in the sea" (as in SKERRIES SH2794), so it does not fit well semantically and the word is not recorded in English until the ME period, with the sense here "rocky outcrop on a hill" being even later. It is a puzzle. Having made a detour around SHILL MOOR NT9415 (which contains either OE *scylf, scelf* "rock, crag" or OE *scylfe, scilfe* "ledge, shelf"), the river bends sharply again below the projecting ridge of SNOUT END NT9516 (ME *snūt(e)* "snout" often used of a projecting hill). RITTO HILL NT9516 with its complex of prehistoric and Roman-period settlements may contain OE *rīþ* "stream" or perhaps a proper name (such as *Ridda(n)* if we can compare it with RITTON NZ0893) and OE *hōh* "spur of a hill" as in CAMBO NZ0285 with the *hill* element added pleonastically. The low hill called MEGGRIM'S KNOWE NT9615, which separates the Breamish from the ROWHOPE BURN (OE *rūh* "rough, uncultivated") certainly seems to contain a personal name, although the settlement, which dates back to the Iron Age, is much older than this name suggests (one *Moregrim* occurs in PASE, associated with the south of England, but the database contains nothing which corresponds exactly). On the other side of the Linhope Burn is GRIEVE'S ASH (also GREAVES ASH) NT9616, a strange name for a hill the top of which is now practically bare, and if Tomlinson is to be relied on (p. 365-6), the name was earlier GREENSHAW HILL (i.e. "green copse"), and Greaves Ash referred to the Iron-Age fortified enclosure. The original name is of course lost, unless it is preserved in the name of DUNMOOR HILL NT9618 (could be **dūnom-mārom* "great fortress", if the second element is not OE *mor* "heath") to the north. However, this is pure conjecture and the present name appears to contain OE *gerēfa* "reeve" and OE *æsc* "ash tree". As "the reeve's ash tree" does not make much sense, it is probably a popular etymology. On the slopes of Dunmoor Hill to the east of DUNMOOR BURN are LONG CRAG NT9617 and CAT CRAG NT9617 (both relatively transparent onomastically), with CUNYAN CRAGS NT9718 just to the east providing a local landmark. The first element of this is obscure. Tomlinson (p. 485) offers "Cunion Crags, or Chieftain Rocks, as they were called by the ancient Britons", but I can find no reconstruction for Brittonic which would support this suggestion. Words beginning in *Cun-* are usually derived from the word for "dog, wolf" (nominative **kū*, genitive **kunos*, W *ci*, plural *cwn*) such as *Cunomaglos* (W *Cynfael*), etc. (it could just possibly be the prefix **kom-* "together").

The farm of LINHOPE NT9616 lies below the ancient monument and probably takes its name from the valley where there are waterfalls, these being BLACK LYNN NT9616 and LINHOPE SPOUT NT9517 (ME *spowte* is occasionally used to indicate a waterfall, e.g. MALLYAN SPOUT in North Yorkshire). HARTSIDE NT9716 is now a farm under Dunmoor Hill, but itself takes its name from HARTSIDE HILL NT9815 (from OE *heorot* "stag" + ME *sīde* "hill" + *hill*, another pleonastic formation), where there are the remains of more prehistoric dwellings. Following the river round the base of Hartside Hill, we come to ALNHAMMOOR NT9715 (*Alnham* + OE *mor* "heath") probably so named to distinguish it from ALNHAM NT9910 "settlement by the Aln", of which more later. The Shank Burn joins the Breamish just downstream, possibly taking its name from EWARTLY SHANK NT9613 (OE *scanca* "shank of a leg" is used in place-names to indicate a projecting hill) which itself takes its name from EWARTLY NT9613 (an alternative name for the farm) meaning "homestead by the water hill / clearing", if we can compare it with EWART NT9631 (*Ewurthe* 1218 = OE *ēa-worþ* "worth on a river"). To the north-west, the RAY CLEUGH (probably OE *rā*, *rāha* "roe deer") runs down from LITTLE DOD NT9514 (see the discussion of THE DODD NY7392 above), with the DOW CLEUGH NT9414 running roughly parallel (probably OE *dā* "doe"). The FORE BURN (could be early MnE *fore* "at the front") is on the other side below NORTH PIKE NT9613 (SOUTH PIKE is in NT9713) and HET HILL NT9614 (*het* is a ME spelling for *head* which fits semantically, but it is pure conjecture). On the other side of Het Hill is COBDEN NT9714, the name of an old cottage just above the waterfalls which seems to contain OE *denu* "valley", but the first element is less straightforward. It could be linked to the *cob* in *cobblestone*, etc. and if so, the existence of a Bronze-Age cairn in the vicinity may be relevant, and the meaning might be "pile of stones valley". The names COBDEN SIKE, COBDEN BURN and COBDEN CLEUGH are probably secondary.

There is another Iron-Age fort — called PRICKLY KNOWE ENCLOSURES NT9814 — on the south-western slopes of PRICKLY KNOWE NT9813. This is clearly OE *cnoll* "hill" and MnE *prickley* may have been motivated by the vegetation, but if so the name must be very modern and most Northumbrian hills are covered in thistles and gorse. Alternatively, the first element may be a popular etymology of Britt. **brigā* "hill" (W *bre*, etc., Gaulish *briga* "hill, fort") and OE *hlāw*, which then underwent the same development as in KIRKLEY NZ1576. The name of the farm CHESTERS NT9814 is likewise not old, but generalized from OE *ceaster*, which was applied first to Roman forts and then to settlements near any fort. The name of the CHESTERS BURN just to the east is easily explicable, sandwiched as it is between Prickly Knowe and the hillfort to the east, i.e. MIDDLE DEAN CAMP NU0014, and the Roman farmsteads on HAYSTACKS HILL NU0015 (the name seems modern and transparent). Perhaps WETHER HILL CAMP NU0015, also known as CORBIE CLEUGH CAMP, helped to motivate the name of the Chesters Burn.

The Breamish turns north to pass between Hartside Hill (see above) and BROUGH LAW NT9916 (reflecting OE *burh* "fort" — there is an Iron-Age hillfort on the summit). BROUGHLOW SLACK NT9915 (the second element derives from ME *slac* "small valley") joins the river to the south west of Brough Law, and provides a spelling of OE *hlāw* "hill" more usually found further south in England. INGRAM GLIDDERS NT9916 probably refers to the scree-like western slopes of the same hill (OE *glidder* "slippery" and OE *gliddrian* "to slide"), and above this the GREENSIDEHILL BURN (GREENSIDEHILL NT9816 is a farm below HUNT LAW NT9816) falls into the Breamish below yet another fort on KNOCK HILL NT9916, which probably reflects Britt. **kunuko-*, another word for "(small) hill" (W *cnwc* "hillock"). Between Knock Hill and EWE HILL NU0016 is SHIVERS CLEUGH NU0016 (the first element possibly a personal name such as *Ceofa* or *Ceofel*). Some readers may have noticed how many hills there are named after ewes and this is so remarkable (there is another EWE HILL NU0015 just to the south) that one may wonder whether we are dealing with a popular etymology. One solution is to derive *ewe* from OE *ēwell* "source of a stream" and the development would then be similar to EWLOE in Flintshire (i.e. *Ewlawe* 1281), with additional vocalization of the [l], of course. Perhaps the duplication of names is a reason why the Iron-Age fort on EWE HILL NU0016 is called REAVELEY HILL CAMP, when the summit of REAVELEY HILL NU0018 is some distance to the north (REAVELEYHILL farm NU0017 is situated on its southern slopes). The early forms of REAVELEY NU0217 (*Reveley* 1268) do not suggest anything other than "the reeve's clearing" (OE *(ge)rēva* "reeve, officer"), the name appearing in secondary names in the vicinity, such as REAVELEY COTTAGE NU0217, REAVELEY BRAES NU0116, REAVELEY FOX COVERT NU0117, and REAVELEY GREENS NU0118, as well as our hill-names discussed above (including GRIEVES ASH). This may be an indication of the importance of the settlement or its owner (we get our modern word *sherriff* from OE *scīr(ge)rēva* "the judicial president of a shire").

The Breamish then reaches the village of INGRAM NU0116 (e.g. *Agreham* 1244, *Angerham* ca 1250) which probably means "the homestead on the meadow" (OE *anger*), the name sometimes being applied erroneously to the valley as a whole. As it was another important settlement, secondary names are well evidenced, including INGRAM FARM NU0116, INGRAM MILL NU0216 and INGRAM HILL NU0115 below TURF KNOWE NU0015 on the MIDDLEDEAN BURN which falls into the Breamish below INGRAM BRIDGE NU0016. The MIDDLE DEAN is the valley (OE *denu*) between the FAWDON DEAN NU0215 and the valley of the Breamish itself. Naturally, one FAWDON BURN is marked as running through Fawdon Dean, but there is another FAWDON BURN which runs through the village of FAWDON itself NU0315 (*Faudon* 1207). The name probably reflects OE *fāh* "multicoloured" plus OE *dūn* "hill", but as with many other *-don* names there is a connection with an Iron-Age hillfort (Britt. **dūnom*), this time on OLD FAWDON HILL

NU0214, and this may be where the name originated. There were also forts on the other side of the burn on CASTLE KNOWE NU0314 (such sites were occasionally called *castle*) and GIBB'S HILL NU0314, so we are spoiled for choice. The first element of this last one may be a common pet form of *Gilbert*, which has been proposed also as the origin of GIBSIDE NZ1858 and GIBSHIEL NY8093, although we should note the widespread use of *gibbet hill* in ME for a hill with a gallows on it. Fortunately for toponymists, EAST HILL NU0316 and WEST HILL NU0214 present no significant problems. Between them are smaller eminences, SNAIL KNOWE NU0215 (OE *snegel* could also mean "slug") and CHAPEL KNOWE NU0215 (we note the remains of a cross marked on the map).

Beyond Ingram Mill, the valley of the Breamish broadens out slightly with HEDDON HILL NU0217 (one of several, all reflecting OE *hēþ* and OE *dūn* "hill") on the northern bank, whose cultivation terraces have attracted some attention, but (perhaps surprisingly for this area) there does not appear to be an old hillfort on the top. To the north of some small expanses of water is BRANDON NU0417, now a farm and the remains of a church, but once a medieval village. The early forms (e.g. *Bremdona* c 1150, *Bromdun* 1236) point to OE **brēmen* "broomy" as first element (the [o:] of OE *brōm* "broom" was affected by the original vowel in the following syllable) with OE *dūn* "hill", so perhaps, given the situation, the name originally referred to a hill nearby (NOVA SCOTIA NU0318 is clearly a relatively recent coining for a hill). What we know for certain is that the name gave rise to secondary toponyms, including BRANDON DEAN NU0418 (the hill to the east has no name but does have the remains of an Iron-Age fort near the wood), BRANDON HILLHEAD NU0517, and BRANDON WHITE HOUSE NU0517 on the Roman road (the so-called Devil's Causeway). BRANTON NU0416 is to the south of the gravel workings (now BRANTON PONDS nature reserve) and the early forms (e.g. *Brementona* ca 1150) differ only from its northern neighbour in suggesting OE *tūn* "farm", rather than OE *dūn* "hill". BRANTON BUILDINGS NU0415 (*buildings* is indicative of a modern name) and BRANTON MIDDLE STEADS NU0414 (OE *stede* "place, building" so may be older) belong here too. One final thought on the Brandon / Branton pair. Nobody seems to have noticed that the first element of the name of the river Breamish appears to be the same as that of the villages in the early attestations, so perhaps we are simply dealing with the hill (fort) and farm on the river **Bremia*.

DUN'S KNOWE NU0315 looks like a knoll or small hill where a fort (Britt. **dūnom*) once stood, but there is nothing marked on the map. Just to the north, COW HILL NU0416 probably reflects OE *cū* "cow", as we would expect, or possibly *col(e)* "charcoal", a hill where charcoal burners worked. GLANTON NU0714 (*Glentendon* 1186, *Glantendon* 1200) is on the western slope of GLANTON HILL NU0614 with GLANTON PYKE NU0614 (the modern spelling is immaterial) just to the west. But the location of Glanton Hill Iron-Age fort is on the northern slopes of the same hill, which leads me to suspect that the name derives from the fort (Britt. **dūnom*) on the slope or bank (Britt. **glandā*, as in *Camboglanna*, the Roman fort at CASTLESTEADS NY5163, and *Glannoventa*, the

Roman fort at RAVENGLASS SD0895, W. *glan* "bank (of a river), hillside") rather than the other etymologies which have been proposed. The first involves OE **glente* "look-out", which at least has the merit of taking the situation near the Roman road into account, but this word is a dubious reconstruction. The second involves Swedish *glanta* "hawk", but this does not seem specific enough for a location name and Scandinavian names are unlikely anyway (see the Introduction). There are two Iron-Age settlements associated with SHAWDON, at SHAWDON WOOD HOUSE NU0815 and SHAWDON HILL FARM NU0813, so one might imagine an etymology involving OE *dūn* "hill" or Britt. **dūnom* "hill-fort", but the old forms (e.g. *Schaheden* 1232) point firmly in the direction of "copse valley". The SHAWDON BURN is in fact a tributary of the River Aln. TITLINGTON NU1015 (*Tedlintona* 1123-8) locates another complex of Iron-Age settlements, but the farm must have been renamed after an Anglian settler and his family (called something like *Titill*, and one person of this name is recorded in PASE). TITLINGTON WOOD NU0916, TITLINGTON PIKE NU0815, and TITLINGTON BURN are all secondary names.

HEDGELEY ('hɪdʒlɪ) is reflected in the name of LOW HEDGELEY FARM NU0717, located near the point where the Roman road crosses the Breamish, with HEDGELEY MOOR NU0519 to the north. The old forms (e.g. *Hiddesleie* ca 1150) point to a clearing belonging to someone called Hiddi who is recorded in LVD. HEDGELEY HALL NU0717 is to the south of the POW BURN, essentially a continuation of the WHAUPMOOR BURN, while POWBURN itself NU0616 is to the south-west on the northern slopes of Glanton Hill. The *pow* element invites comparison with the Cumbrian POW BECK NX9712, a stream that falls into the Irish Sea at Whitehaven, and the two other Cumbrian examples of POW BECK (NY2424, NY3949) all reflecting ME *poll* "pool" with vocalization of [l] to [w] which we also find in local realizations of MnE *old*, e.g. [awd]. This ME *poll* is found in stream names, most frequently in Cumbria, and there are also reflexes with the old [l] intact, as in the name of the POLTROSS BURN, a small tributary of the Irthing which forms the boundary between Northumberland and Cumbria (which, following Ekwall, we could derive from this *poll* and a Britt. **trāns-* "across", as in TRAWSCOED SO0834 in Powys). ME *poll*, and MnE *pool* of course, may reflect OE *pōl* "pool", which has cognates in West Germanic (OHG *phuol*, MLG *pōl*, MDu *poel*) but no further certain connections. There is also Welsh *pwll* "pool" (Ekwall's suggestion that it formerly meant "stream" cannot be substantiated), OC *pol* which glosses Latin *puteus* "well, pond" (it is one of the words along with *Tre* and *Pen* which commonly distinguish Cornish names) and Breton *poull*, which suggests that we would be justified in reconstructing a common Brittonic word. It is always possible that *poll* was borrowed into late Brittonic from OE, but then one would have to explain why the place-names reflecting it are concentrated in formerly Brittonic-speaking areas. The pool in question might have been one of the small areas of water around the area where the Pow Burn falls into the Breamish. On the hill to the east of Powburn is CRAWLEY FARM NU0616 (*Crawelawe* 1225) or "crow hill" (OE *crāwe*), which also gives its name to CRAWLEY DEAN NU0616. The MUGSDEN BURN NU0717 flows through a valley to the north of Hedgeley Hall in which inferior sheep (Nthb. *mugs*) may have been kept. The location of a hydraulic ram in the valley is an almost poetic coincidence.

GALLOW LAW NU0618 is a small hill north of the river reflecting OE *gealga* "gallows", so this was probably a hill where local criminals were despatched. It is not an uncommon name, and the old building on the northern slopes probably takes its name from the hill. BEANLEY NU0818 (*Benelega* ca 1150) to the south of the river is named after a clearing where beans (OE *bēan*) were cultivated, an important part of the diet in medieval England. The settlement was probably more important than the name suggests, as it has given its name to BEANLEY MOOR NU1018, BEANLEY PLANTATION NU0917, BEANLEY MOSS NU0917, as well as BEANLEY HALL NU0818 near the site of the medieval village. There are two Iron-Age settlements on the hills overlooking the valley, and one to the north-east called THE RINGSES NU0918, which has a double plural (*rings*, already plural, acquires another plural suffix *-es*, like LEAZES, from OE *lēahas* plus *-es*), and is a name applied to a number of circular hillforts in Northumberland. The lower reaches of the Breamish have been planted with woodland: BLACKBRIDGE PLANTATION NU0718, PRINGLE'S WOOD NU0718, WATERSIDE PLANTATION NU0719 (planted between 1850 and 1900) and HALL WOOD NU0820 all have a modern ring about them (and indeed the first two were planted after 1950, the latter, clearly connected with HAREHOPE HALL NU0820, in the second half of the 19th century). HAREHOPE NU0920 (*Harop* 1185) was a small farm (now residential) overlooking the HAREHOPE BURN, which also shares its name with HAREHOPE HILL NU0820 above. The first element looks like OE *hara* "hare", but could just be *hār* in its derived meaning of "boundary"; *-hope* could refer either to the valley of the Harehope Burn or to the farm itself, i.e. in the meaning "piece of land". The Harehope Burn flows down from Hepburn Moor (see below), over DOVE CRAGS NU0922 (see other similar names elsewhere in the county) and then over CORBIE CRAGS NU0821 (ME *corby* "raven, crow") and GREY MARE'S CRAG NU0821 (we have seen that these names are often associated with boundary stones) before joining the STOCK BROOK (OE *brōc* "stream") below TICK LAW NU0821 (OE *ticia*, probably an error for **ticca* = NHG *Zecke* "tick"). We should note that *brook* is not the typical word for a stream in Northumberland (an old name for the Forest Burn in the Coquet Valley — *Foresty brooke* 1577 — is the only other certain example): OE *burna* and *well(a)* are used in almost all cases.

OLD BEWICK ['bju:ɪk] NU0621 (*Beuuiche* ca 1136) is the old bee farm (OE *bēo*). Honey was an important product in early medieval England, both for its then almost unique sweetening properties and as the starting point for mead. One can imagine that the hives were able to thrive on the south-west facing slopes above the river. NEW BEWICK NU0620 is on the other bank. The early 19th-century BEWICK BRIDGE NU0522 is downstream, and BEWICK FOLLY NU0622 to the north of the KIRK BURN, a stream named after the 12th-century Holy Trinity

Church NU0622. Of course, as mentioned above in connection with other Northumberland follies, *folly* probably does not signify a curious building, but more likely a clump of trees on a hill, and has given its name to FOLLY WOOD NU0722, FOLLY BANKS NU0622 and FOLLY BURN, which in turn is the origin of FOLLYBURN WOOD NU0623. BEWICK HILL NU0721 has a number of hill-forts and cup-and-ring marked stones, and behind is BEWICK MOOR NU0922 (both of these are secondary names) with CASTLE HILL NU0922, probably a modern name motivated by the prehistoric hut circle to be found there. Behind Old Bewick, a path leads to BLAWEARIE NU0822, an isolated, now disused farmhouse on the edge of Bewick Moor. The wind does indeed "blow weary" up here, but this often cited derivation seems to rely too much on "what the name sounds like in Modern English" and to ignore the other places with the same name in the borders (including NU0639, NY5917 and NT4416). If it is a popular etymology, the base could be Brittonic, and BLAENWERN, of which there are three in Cardiganshire (SN6253, SN4256, SN3446), one in Carmarthenshire (SN5542) and one in Pembrokeshire (SN0327), as well as BLANERNE (NT8356) in the Scottish Borders, which can be explained by loss of [w] in English (as in *Berwick*), offers a possible parallel. BLAENWERN can be derived from **blanyā-wernā* "edge of the moor" which would fit the locations in question better. There is still the problem of the loss of [n], however, but popular etymologies sometimes confound regular sound-changes. There is also a CASTLEWEARY NT4003 in the borders, which supports the analysis of *wearie / weary* as a discrete element, but unfortunately does not make its origin any clearer.

Here our journey down the Breamish ends. As the old folk-rhyme says

> Foot of Breamish and head of Till
> Meet together at Berwick Mill.

And this also concludes our tour of the place-names of the western uplands of Northumberland, which I hope you have enjoyed. The other books planned for the series will deal with the Tyne and Allen valleys, Tyneside and the Northumberland plain.

Further Reading

Cameron, Kenneth, 1996. *English Place Names*. New edition. London: Batsford.

Crystal, David, 2004. *The Stories of English*. London: Allen Lane / Penguin.

Delamarre, Xavier, 2003. *Dictionnaire de la langue gauloise*. 2nd edition. Paris: éditions errance.

Dixon, David Dippie, 1905. *Upper Coquetdale, Northumberland. Its History, Traditions, Folk=lore and Scenery*. Newcastle upon Tyne: Robert Redpath.

Ekwall, Eilert, 1928. *English River Names*. Oxford: Clarendon Press.

Ekwall, Eilert, 1960. *The concise Oxford dictionary of English place-names*. Oxford: Clarendon Press. (4th edition, first published 1936.)

Heslop, O., 1892-4. *Northumberland Words*. 2 volumes. London: Kegan Paul.

Jackson, K. H., 1953. *Language and History in Early Britain*. Edinburgh: EUP.

Mawer, A., 1920. *The Place-Names of Northumberland and Durham*. Cambridge: Cambridge University Press. (Bibliolife Reprint, n.d.)

Mills, A. D., 1991. *Oxford Dictionary of British Place-Names*. Oxford: OUP.

Nicolaisen, W. F. H., 1976. *Scottish Place-names*. London: Batsford.

Owen, Hywel Wyn and Richard Morgan, 2007. *Dictionary of the Place-Names of Wales*. Llandysul: Gomer Press.

Price, Glanville, (ed.) 1998. *Encyclopedia of the Languages of Europe*. Oxford: Blackwell.

Reaney, P. H., 1958. *A Dictionary of English Surnames*. Oxford: University Press. (Revised third edition with corrections and additions by R.M. Wilson, 1995, 2007.)

Rivet, A. L. F. and Colin Smith, 1981. *The Place-Names of Roman Britain*. London: Book Club Associates. (Originally published London: Batsford, 1979.)

Sims-Williams, Patrick. *Ancient Celtic Place-Names in Europe and Asia Minor*. Oxford: Blackwell. (Publications of the Philological Society, 39.)

Tomlinson, W. W. 1888. *The Comprehensive Guide to the County of Northumberland*. Newcastle upn Tyne: Scott. (Republished Durham: David Books, 1985.)

Watson, Godfrey, 1970. *Northumberland Place Names. Goodwife Hot & others*. (Morpeth: Sandhill Press)

Watson, W. J., 1926. *The History of the Celtic Place-Names of Scotland*. Edinburgh and London: W. Blackwood. (Republished Edinburgh: Birlinn, 2004).

Watts, Victor, 2004. *The Cambridge Dictionary of English Place-Names*. Cambridge: Cambridge University Press. (Paperback edition 2010.)

Some Technical Terms

Anglian: the name for the northern Old English people who settled the part of Britain north of the Humber (later Northumbria). This name also applies to their dialect.

Brittonic: a name for the largely reconstructed P-Celtic language of Britain before the Roman conquest (1st century AD). After the English settlement (from the 5th century AD) Brittonic and its descendants are gradually replaced by English.

Celtic: the branch of the Indo-European languages which includes Welsh, Irish, Cornish and Breton (= Insular Celtic) as well as Gaulish and Celtiberian (= Continental Celtic). The Celtic languages are also grouped according to their treatment of the Indo-European sound we find in Latin *quinque* "five": P-Celtic has [p] (e.g. Welsh *pump* [pɪmp] "five") and Q-Celtic has [k] (e.g. Irish *coic* "five").

Cognate: cognate forms derive from the same source, they are "born together". So English *father*, German *Vater*, Latin *pater*, Irish *athair*, etc. are all cognates.

Dative: the case of a noun with a general meaning "to". Relics of the OE dative, used as a locative, appear in some place names.

Diminutive: a word or part of a word indicating a little thing, e.g. *-let* in *piglet* "little pig".

Diphthong: a movement from one vowel to another within a single syllable. So [aɪ] is a diphthong, whereas [a] and [ɪ] are monophthongs.

Genitive: the case of a noun typically indicating possession. In the phrase *John's book*, *John's* is in the genitive, one of the few relics of the case system remaining in English.

Germanic: the branch of the Indo-European languages which includes English, German and Dutch (West Germanic), Icelandic, Faroese, Norwegian, Swedish and Danish (North Germanic or Scandinavian), and Gothic (East Germanic).

Great Vowel Shift: the English sound change whereby the long vowels [i:] and [u:] became diphthongs [aɪ] and [aʊ] (so OE *fīf* [fi:f] changed to MnE *five* [faɪv] and OE *lūs* [lu:s] changed to MnE *louse* [laʊs]) and other long vowels were came to be pronounced with the tongue closer to the roof of the mouth (e.g. [e:] became [i:], as in MnE *cheese*, the spelling often lagging behind the pronunciation).

Haplology: the deletion of one of two identical or similar contiguous segments — *haplology* would become *haplogy* via haplology.

Indo-European: the large family of languages which includes most European languages as well as Iranian and major languages of the Indian sub-continent (see Price 1998).

Locative: a case of a noun used to indicate where something is located. There was no longer a distinct form for the locative in the Germanic languages, so the dative was used (as in OE *ācum* "by the oak trees")

Metathesis: where two sounds change places (so OE *brid* becomes MnE *bird*). *Metathesis* would become *metasethis via metathesis.

Motivation: the reason for a place-name (e.g. TOD LAW is so named because foxes were found there).

Oblique case: in inflecting languages (such as Old English, Latin or Brittonic), a collective term for any case other than the nominative. The accusative, genitive, dative and locative are examples of oblique cases.

Onomastics: the study of names. An onomastic tale is a folk-tale used to explain a place-name.

Palatalization: occurs when the articulation of a sound moves towards the hard palate. Used specifically in Old English to indicate the change of velar [k] to [tʃ] (as in *kirk* > *church*).

Pleonasm: a pleonastic formation (see below).

Pleonastic: a pleonastic formation is when a more familiar element is added to clarify a less familiar element. This process can occur repeatedly. So *Pen* might become *Pen Hill*, because *Pen* (W *pen* "head, top") was not well understood. When *Pen Hill* became *Pendle*, because the first syllable was stressed and the second was weakened and a [d] was added, as happened occasionally in this phonetic context, *Pendle* was felt to be unfamiliar or opaque and another *Hill* was added. So *Pendle Hill* actually means "hill hill hill". If I had to choose the most important process of place-name formation in Northumberland, this would be it.

Popular Etymology: an unfamiliar word is replaced by a more familiar one (e.g. foreign-sounding *asparagus* is replaced by native-sounding *sparrow grass*).

Prefix: an element added onto the beginning of a word (in *delouse*, *de-* is a prefix).

Root: one of the elements of an inflected word in the older Indo-European languages. All inflected words had a root, most had an inflectional class marker, and the ending showed how the word fitted into the sentence. An example is the Latin word for "good" *bon-u-s*, where *bon-* is the root, *-u-* is the class marker and *-s* is the ending.

Semantics: the branch of linguistics devoted to the study of meaning, so *semantically* "from the point of view of meaning".

Suffix: an element added on to the end of a word (e.g. in *reading*, *-ing* is a suffix).

Toponymy: the study of place-names, or toponyms (so a toponymist is someone who studies place-names).

Vocalization: occurs when a consonant becomes a vowel. This often happens with [r] and [l], so there is no distinction in standard English between *father* and *farther*, and standard English *all* becomes Nthb. [aː].

Index

Abbey Rigg NY8487	21
Acomb NY9366	33
Addycombe NU0502	76
Ainsey Knowe NT9115	94
Akeld NT9529	88
Akenshaw Burn	49
Akenshawburn NY6089	49
Allerdene NU0201	75
Allerhope Burn	79
Allerybank NY7481	45
Allgood Farm NY8574	37
Alnhammoor NT9715	95
Alwin	65, 66
Alwinton NT9206	65
Angryhaugh NT9205	10, 65
Anick Grange NY9565	32
Anick NY9565	32
Anton Hill NY8580	43
Archy's Rigg NY7083	46
Arks Edge NT7107	9
Arks NT7108	9
Ash, The Ash NY8177	40
Ashtrees NY8395	16
Ashy Bog NY7595	55
Aydon Castle NZ0066 NE45 5PJ	32
Babswood Kirk NT7402	13
Back Burn	65, 75
Bagraw Farm NY8596	17, 20
Bakethin NY6491	50
Bakethin Reservoir NY6391	50
Bamburgh NU1834	13
Bank Foot NY9565	32
Bankhead NY8479	42
Barnes, The Barnes NY8477	41
Barrasford NY9173	28
Barrow Burn	66
Barrow Hill NT9004	66
Barrow Law NT8611	69
Barrow NT9106	66
Barrow Scar	66
Barrowburn NT8610	69
Batailshiel Haugh NT8810	81
Bateinghope	12
Bavington Hall NE19 2BA	29
Beacon, The Beacon NT9500	63
Beanley NU0818	99
Beaufront Castle NY9665	32
Beaufront Hill Head NY9666	32
Beaufront Red House NY9765	32
Beaufront Wood Head Farm NY9566	32
Beaumont House NY9572	30
Beef Stand NT8213	69
Beefstand Hill NT8114	69
Beefstand Hill NT8214	71
Bell Crags NY7773	38
Bell Hill NT8410	58, 70
Bell, The Bell NT9028	89
Bell's Braes NY6871	38, 39
Bellcrag Flow NY7773	38
Belling Burn	53
Belling Rigg NY7890	55, 58
Belling, The Belling NY6988	53, 58
Bellingburn Head NY6991	53
Bellingham NY8483	58
Bellion NZ0889	58
Bells Moor NY6095	50
Bellsburnfoot NY6194	50
Bellshiel Crag	15
Bellshiel Law NT8101	15
Bellshiel NY8199 NE19 1TE	14, 58
Bellshiel Pit Cottage	14
Bendor NT9629	88
Bennettsfield NY8595	17
Bennett's Rigg NY8596	17
Bent House NY7785 NE48 1LD	45
Beukley Covert	30
Beukley NY9870	30
Bewick Folly NU0622	99
Bewick Hill NU0721	99
Bewshaugh Cottages NY6391	50
Bickerton NT9900	74
Bieldy Pike NU0604	77
Billerley NY8479	42
Billsmoor Park	60
Billsmoorfoot NY9497	61
Bimmerhill NY8086	57
Bingfield Combe NY9872	30
Bingfield East Quarter NY9872	30
Bingfield East Side NY9873	30
Bingfield NY9772	30
Binky Burn NY6684	47
Birdhope Craig Hall	15
Birk Hill NY7876	40
Birkhill NY8595	17
Birks NY7784	44
Birky Grain	49
Birtley NY8778	27
Birtley Shields NY8779	27
Bizzle, The Bizzle NT8922	91
Black Bog Burn	27
Black Brae NT9800	74
Black Burn	55
Black Burn NT8202	15
Black Carts NY8871	9
Black Chirnells NU0303	76
Black Cleugh NT6905	12
Black Cleugh NT9215	94
Black Fell	39

Black Fell NY6093	50	Boughthill NY7886	57
Black Fell NY7073	38, 39	Bowent	89
Black Hag	90	Bower NY7583	45
Black Hill NY5784	48	Bowershield NY9494	60
Black Hill NY8195	16	Boweshill NY8085	57
Black Hill NY9097	60	Bowmont	89
Black Hill NY9297	60	Bran's Walls NY6697	51
Black Hill NY9767	32	Brandon NU0417	97
Black Knowe NT8209	68	Brands Hill NT8723	85
Black Knowe NY5891	50	Branton NU0517	97
Black Knowe NY6481	45	Braydon Cleugh NT8821	91
Black Knowe NY6780	44	Breadless	13
Black Law NY8073	37	Breamish	93
Black Linn NT8409	68	Breamish Head NT9018	93
Black Lynn NT9616	95	Bremenium NY8398	15
Black Middens Bastle NY7790	55	Bridge End NY9166	32
Black Stitchel NY9098	18	Bridge End NY9564	32
Blacka Burn	40	Bridge House NY8279	42
Blackaburn NY7977	40	Bridgeford NY8582	43
Blackblakehope NY7599	14	Brieredge NY8083	43
Blackbridge Plantation NU0718	99	Brigantium NY8398	15
Blackburn Common	55	Brigg, The Brigg NY8989	21
Blackburnhead NY7793	55	Brinkburn NZ1198	78
Blackcleugh Burn	47	Broad Moss NT9621	86
Blackhag Sike	54	Broadhope Hill NT9323	86
Blackhaggs Rigg NT8724	91	Broadpool Common	38
Blackhill Farm NY8876	28	Broadside Law NT8211	70
Blackkip NT7904	14	Broadstruthers Burn	85
Blackman's Law NY7498	17, 54	Brockley Park NY9697	62
Blackseat Hill NT9422	86	Broom Park Farm NY9266	34
Blackwool Law NY8098	15	Broomhill NY9086	24
Blakehope Burn	14	Broomhope Mill NY8783	24
Blakehope Nick NY7918	14	Broomhope NY9883	24
Blakehope NY8594	20	Broomy Cleugh NT9316	94
Blakehopeburnhaugh NT7800	14	Broomy Hill NY6385	47
Blakelaw NY9286	23	Broomylinn NY6384	47
Blakelaw NZ2166	17	Brough Law NT9916	95
Blakeman's Law NY8795	17	Broughlaw Slack NT9915	96
Blawearie NU0822	100	Brown Law NT8305	67
Blaxter Cottages NY9390	19	Brown's Law NT9727	84
Blaxter Lough NY9398	19	Brownchesters NY8892	18
Blaxter Quarry NY9390	19	Brownhart Law NT7809	71
Bleaklaw NY8475	37	Brownknowe NY7986	57
Blindburn NT8210	70	Brownsleazes NY8379	42
Blindburn NY8678	27	Brunton Bank NY9369	35
Blinkbonny Cottage NY8380	42	Brunton House NY9269	35
Bloody Bush NY5790	49	Brunton Turret NY9269	35
Bloodybush Edge NT9014	80	Buck Bog Sike	39
Blue Chirnells NU0303	76	Buck Fell NY5990	49
Blue Hemmel Sike NY7475	39	Buckham's Walls Burn NT7911	71
Bluesteel Rigg NY7185	46	Bucklake Sike NY7094	54
Bluestone Edge NT8602	67	Buckside Knowe NY5992	49
Boat Farm NY8482	43	Bullcrag Edge / Peninsula NY6786	47
Boe Rigg NY8085	57	Burdhope, Birdhope NY8198	15
Bog Shield NY8979	27	Burdhopecrag	15
Bogg, The Bogg NY8892	18	Burdon Side NY8090	56
Bolts Law NY6981	45	Burgh Hill NU0200	74
Bough Law NT8412	69	Burmoor NY8277	41

Burn Grange NY7294	54	Cat Loup NT9421	86
Burn House NY8673	37	Catcherside NY9987	13
Burnmouth Cottages NY8974	28	Catcleuch Hill NT7406	10
Burnmouth NY7988	56	Catcleuch NT6806	10
Burnt Tom Crags NY5981	47	Catcleuch Shin NT6806	10
Burnt Tom NY6286	47	Catcleugh Farm NT7403	12
Burradon NT9806	73	Catcleugh House NT7403 NE19 1TX	13
Bushman's Crag NT8403	67	Catcleugh Reservoir	12
Buteland Fell NY8882	25	Catless NY8375	37
Buteland NY8781 NE48 2EX	25	Catreen NY8878	27
Butt Hill NY7477	40	Cats Elbow NY9281	25
Butthill Sike	40	Catterick	9
Bygate Hall Cottages NT8606	68	Chairford Bridge NY9086	22
Bygate Hall NT8505	68	Chantry Farm NY9866	31
Byreshaw Hill NY7672	38	Chapel Knowe NU0215	97
Byreshield Grains NY6883	46	Charity Hall NT9604	73
Byreshield Hill NY6783	46	Charlton Burn	57
Byrness Hill NT7703	13	Charlton NY8084	57
Byrness, The Byrness NT7602 NE19 1TR	13	Chattlehope Burn	12, 13
Cadger Bog NT7301	13	Chattlehope Crag NT7302	13
Cadger Ford NY7683	44	Chattlehope Farm NT7302 NE19 1TY	13
Cadgerford NY7171	38	Chatton NU0528	9
Cairnglastenhope NY7580	44	Cheese Sike	52
Caistron NT9901	74	Cheshill Ways NY9774	28
Calf Lee NT9005	66	Chester Hope NZ0299	74
Callaly NU0509	64	Chesterhope NY8985	24
Caller Cleughs NY6797	51	Chesters NT9814	94
Cambo NZ0285	12	Cheviot, The Cheviot	82
Camp Hill NY9093	59	Chew Green NT7808	66
Camp Hill NY9176	26	Chipchase Mill NY8874	28
Campville NT9402	63	Chipchase NY8875 NE48 3NT	28
Caplestone Fell	49	Chipchase Strothers NY8874 NE48 3PE	28
Carey Burn	85	Chirdon NY7683	44
Carlcroft NT8311	70	Chirdonhead NY7181	45
Carling Crags NT9524	85	Chirnells NU0302	76
Carr Hill NY9567	34	Chiselways Farm	28
Carrick Heights NY9096	59	Chisholm's Moss NY7781	44
Carrow Rigg NY9496	60	Chollerford NY9171 NE46 4EW	35
Carry Burn	26	Chollerton NY9372	35
Carrycoats Hall NY9279	26	Christy's Crags NY6882	45
Carshope NT8411	70	Cilurnum NY9170	34
Carter Bar NT6906	9	Clay Walls NY9879	29
Carter Fell	9	Clemy's Cairn NT8800	64
Carter Pike NT6904	9	Clennell Hall NT9307	79
Carterhouse NT6707	9, 10	Clennell Hill NT9308	79
Carterside NU0400	77	Cleughbrae NY8396	16
Cartington NU0304	75	Click'em In NZ0072	31
Carts, The Carts NY8373	37	Clifton Rigg NT8606	68
Castle Burn	37	Clock's Cleugh	45
Castle Dean NY8673	36	Closehead NY9093	59
Castle Hill NU0922	100	Closehill NY8185	57
Castle Hills NY9102	79	Cloven Crag NY6790	53
Castle Knowe NU0314	97	Cloven Crag NY9597	61
Castle Lane NY8673	36	Coal Burn	27, 39
Cat Cairn NY6192	50	Coal Grain	49
Cat Cleuch NT7403	12	Coalcleugh NY9774	38
Cat Cleugh NT9215	94	Cobden NT9714	95
Cat Crag NT9617	94	Cock Play NY9082	24

Cock Ridge NY8790	21
Cockshaw Sike	86
Codlaw Dene NY9468	34
Codlaw Hill NY9468	34
Cold Law NT9203	65
Cold Law NT9523	86
Cold Law NY9285	23
Coldberry Hill NT9627	88
Coldburn NT8924	91
Coldcotes NY7675	39
Coldgate Mill NT9925	84
Coldlaw Burn	65
Coldlaw Cairn NT9118	93
Coldtown NY8988	22
Coldwell NY9073	36
Coldwell Strand NT9018	93
College Burn	89
Colt Crag Reservoir NY9378	26
Colwell NY9575	29
Colwellhill NY9194	59
Comb Bastle	5
Comb Fell NT9118	86
Comb Hill NY7692	55
Comb Hill NY9181	25
Comb NY7690	55
Combs Burn	61
Comogan Farm NY8776	28
Conshield NY8575	37
Coomsdon Burn	12
Copper Snout NT8809	81
Coquet	66
Coquet Head NT7808	72
Coquet Island NT9204	66
Corbie Castle	55
Corbie Cleugh Camp	96
Corbie Crag NT8623	91
Corbie Crags NU0821	99
Corbridge NY9964	32
Corby Linn NT8506	68
Corby Pike NT8401	15
Coronation Wood	85
Corsenside Common NY8688	21
Corsenside NY8989	21
Cote Hill NT9900	74
Cottonshope Burn	14
Cottonshope Farm NT7904 NE19 1TF	14
Cottonshope Head NT8006	14
Cottonshopeburnfoot NT7801 NE19 1TF	14
Countess Park NY9780	26
Cow Hill NU0416	97
Cowden NY9179	27
Cowstand Burn NY9078	27
Cowstand Hill NY9381	25
Crag Farm NY8885	24
Crag Head NY6194	50
Crag House NY9269	34
Craig NY9399	61
Craigshield NY8077	40
Cranester Bog NY9586	23
Crawley Farm NE66 4JA NU0616	98
Crigdon Hill NT8605	67
Croft Sike	68
Crook Burn	36
Crookbank NY7876	39
Cross Cleugh NT6905	12
Cross Law NY8689	22
Crossridge NY8377	42
Crow Bridge NY8888	22
Crowdie Law NY7983	44
Crutch, The Crutch NT9800	73
Cuddy's Cleugh NY9487	23
Cunyan Crags NT9718	94
Cushat Law NT9213	81
Daw's Hill NT9400	61
Dally Castle NY7784 NE48 1LH	44
Darden Pike NY9695	62
Darden Rigg NY9896	62
Dargues Burn	20
Dargues Hope NY8493	20
Dargues NY8693	20
Darney Crag NY9187	23
Darney Hall NY9187	23
Darney Quarry NY9188	23
Davidson's Linn NT8815	80
Davyshiel Common NY8897	18
Davyshiel Common NY8997	59
Davyshiel Farm NY8996	59
Daw's Moss NT9300	61
Deadwater Fell NY6297	51
Deadwater Moor NY6398	51
Deadwater NY6096	51
Dean Burn	41
Debdon Lake NU0602	76
Debdon NU0604	77
Deep Sike	54
Deer Play NY8490	21
Deerbush Hill NT8308	68
Dere Street	9
Devil's Elbow	65
Dings Rigg NY7084	46
Dinley Burn	28
Dinley Hill NY8877	28
Dinmont Lairs NY6290	50
Divethill NY9879	29
Dod, The Dod NY9187	23
Dodd, The Dodd NT9209	79
Dodd, The Dodd NY7392	54
Donkleywood NY7486	54
Dough Crag NY9795	62
Dour Hill NT7902	14
Dove Crags NU0922	99
Dove Sike NY6897	51
Dow Cleugh NT9414	95
Drake Stone NT9204	65
Drakestone Burn	65
Drove Rigg NY7085	46

Drowned Sike	49	Fairshaw Farm NY8873	36
Dry Burn NY9380	25	Fairspring Farm NY9974	29
Dryhope NT9211	80	Fairwood Fell NT7307	10
Dudlees NT8600	64	Fallow Knowes NT8507	68
Dues Hill NT9500	63	Fallowfield NY9268	34
Dueshill NT9601	63	Falstone NY7287	53
Dumb Hope NT8509	68	Fawcett Hill Cottage NY9767	32
Dumbhope Law NT8505	68	Fawcett NY9676	29
Dumhope Burn	68	Fawdon Dean NU0215	96
Dun Moss NT7007	10	Fawdon Hill NY8993	18, 59
Dun's Knowe NU0315	97	Fawdon NU0315	96
Dun's Pike NY7782	44	Fawhope NT7409	10
Dun's Pike NY7881	43	Felecia Crags NY7277	40
Dunmoor Hill NT9618	94	Fell House NT9576	29
Dunn's Farm NY9396	60	Felton Hill NY9180	25
Dunns Houses NY8692 NE19 1LB	20	Fenwickfield NY8573	37
Dunsdale Crag NT8923	91	Fern Hill NY9567	34
Dunshaw Farm NY9274	28	Ferny Knowe NY6289	50
Dunshiel NY9294	59	Fiddler's Wood NY8898	18
Duntae Edge NT6402	10	Fieldhead NY8086	57
Dunterley NY8283 NE48 2JZ	43	Five Kings NT9500	63
Dyke Head NY8889	21	Fleehope NT8823	91
Dykehead NY8398	15	Flesh Shank NT8910	80
Ealingham NY8480	42	Flothers, The Flothers NY7076	39
Eals Clough NY7483	45	Flotterton NY7902	75
Eals NY8482	43	Folly Moss NY9377	29
Earl's Seat NY7192	54	Folly Wood NU0722	100
Earle Mill NT9926	84	Folly, The Folly NY9294	59
Earle NT9826	83	Fore Burn	95
East Hepple NT9800	73	Forking Sike	49
East Highridge NY8280	42	Foul Whasle	71
East Hill NU0316	97	Foulplay Knowe NT8900	64
East Wilkwood NT8902	67	Fourlaws NY9082 NE48 2EY	24
East Woodburn NY9086	22	Fourlawshill Top NY9083	24
Easter Dean	85	Fox Covert NY9973	30
Easter Tor NT 9128	89	Foxhole Sike	49
Elishaw NY8695 NE19 1JH	13, 17	Foxton NT9605	73
Elliot's Pike NY5987	49	Fredden Hill NT9526	84
Ellis Crag NT7401	13	Fulhope Edge NT8209	71
Elsdon Gate NY9293	59	Fulhope NT8110	71
Elsdon NY9393 NE19 1AA	60	Gains Law NT9528	88
Emblehope NY7494	54	Gallow Law NU0618	99
Emlope Crags NY7495	54	Gallow Law NY7582	44
Emmethaugh NY6987	46	Gamelspath	66
Erring Burn	30	Garleigh Moor NZ0699	77
Errington Hill Head NY9669	30	Garret Hot NY8681	26
Errington NY9571	30	Garret Lee NZ1096	78
Errington Red House NY9771	30	Garretshiels NY8693	20
Esp Hill NY7979	42	Gatehouse NY7888	56
Esp Mill NY8279	42	Geordy's Knowe NY6287	49
Evistones NY8396	16	Geordy's Pike NY5985	48
Ewartly Shank NT9613	95	Geordy's Sike	49
Ewe Hill NU0015	95	Ghyllheugh NZ1397	78
Ewe Hill NU0016	95	Gibb's Hill NU0314	97
Ewe Hill NY6797	51	Gibshiel NY8093	56
Ewe Lairs NY6591	52	Gill Hassock NY6184	47
Fairloans NT7508	10	Gill Pike NY6183	47
Fairneycleugh NY9194	59	Gilliehill Clints NY7790	55

Cheviot Hills and Dales

Gills Law NT9409	79
Gimmer Knowe NT8407	67
Gimmerstone NY7991	56
Girdle Fell NT7001	13
Girdle Stone	51
Girsonfield NY8993	18
Glanton Hill NU0614	97
Gleadheugh Wood NZ0999	78
Gleadscleugh NT9529	88
Gleedlee NY7789	55
Glen Ridley NY8676	41
Glen, River Glen	88
Glendhu Hill NY5686	49
Glitteringstone U0303	75
Goatstones NY8474	37
Gofton NY8375	37
Gold Island NY8677	27
Goodwife Hot NY8778	26
Gorless NY8293	20
Gowanburn NY6491	52
Gowk Hill NU0904	77
Gowk Hill NY9585	23
Graham's Cleugh	67
Grains Burn	49
Grasslees NY9597	61
Great Bavington NY9880	29
Great Buckster NY7172	39
Great Dour NT7903	14
Great Hetha NT8827	90
Great Lonborough NY8273	37
Great Moor NY8590	21
Great Swinburne NY9375	28
Great Tosson NU0200	74
Great Wanney Crag NY9383	25
Great Watch Hill NY7075	38
Great Whittington NZ0070	31
Greave's Ash / Grieve's Ash NT9616	94
Green Burn	53
Green Eyes Crags NY7388	53
Green Side NT9007	81
Green Swangs NY8086	57
Greenchesters NY8794	17
Greenhaugh NY7987	57
Greenhaugh NY8572	36
Greenhead NY8383	21
Greens' Gears NY5985	48
Greenshaw Sike	49
Greensidehill NT9816	96
Greenwood Law NY8900	64
Greenwood Law NY8999	18
Grey Mare's Crag NU0821	99
Grey Mare's Crags NY6182	47
Grey Stone NY8290	21
Greyhound Law NT7606	10
Greymare Rigg NY8998	18
Greystead NY7785 NE48 1LE	45
Grimping Haugh NT9825	84
Grindon Green NY7273	39
Grindstone Law NZ0073	31
Grindstone Sike NY9187	23
Grindstonelaw Farm NZ0073	31
Grottington Farm NY9769 NE19 2LB	30
Gunnerton Burn	27
Gunnerton NY9075	28
Habitancum NY8986	22
Hag Sike	38
Haggie Knowe NT6301	10
Haggle Rigg NY8374	37
Haining Head NY9292	18, 59
Haining, The Haining NY7575	39
Haining, The Haining NY9292	59
Hainingrigg NY8484	21
Halfway House NY9367	34
Hall Barns NY8773	36
Hall Wood NU0820	99
Hallington NY9895	29
Hallington Reservoir	29
Hallyards NY9086	23
Hampstead NY9865	32
Harbottle NT9304	64
Harden Edge NT7807	72
Hardriding NY7563	61
Hare Law NT9026	90
Hare Rigg NY7084	46
Harehaugh Hill NY9699	62
Harehope Hill NT9528	88
Harehope NU0920	99
Hareshaw Cleugh	94
Hareshaw Head NY8588 NE48 2JB	21
Hareshaw House NY8487	21
Hareshaw Linn NY8485	21
Hareshaw NY8488	21
Harewalls NY9628	23
Harpath Sike	93
Harper Burn NY7093	54
Harper Crag NY7093	54
Harrowbog NT8927	89
Harthope Burn	84, 86
Harthope Linn NT9220	86
Hartside NT9815	94
Hartside NY9287	23
Haughton Castle NY9272	35
Haughton Common	36
Haughton Mains NY9271	35
Haughton Strother NY9273	35
Hawk Hirst NY8079	42
Hawk Knowe NY7294	54
Hawkhirst Cottage NY6688	48
Hawkhirst NY6689	48
Hawkhope Hill NY7287	53
Hawkhope NY7188	53
Hawsen Burn	86
Haystacks Hill NU0015	95
Hazely Burn	85
Headshope NY9399	61
Heart's Toe, The NT7606	10

Heatheridge NY8972	36	Holystone Common	64
Heatherwick NY8992	18	Holystone Grange NT9600	63
Heathery Hall NY7889	56	Holystone NY9502	63
Heathery Hill NY8993	18	Hope Sike	38
Heddon Hill NU0217	97	Hopealone NY7371	38
Hedgehope Hill NT9419	86	Hopefoot NY8895	59
Hedgeley Hall NE66 0HZ NU0717	98	Hopehead NY8996	59
Heely Dod NY9298	60	Hopeshield Burn	37
Heigh NT9111	80	Horneystead NY8177	40
Hen Hole NT8820	92	Horsdon Hill NT9827	83
Hepden Burn	69	Horsley NY8496 NE19 1TA	16
Hepple NT9800	73	Hosedon Burn	65
Hepple Whitefield NY9899	62	Hot Heads NY9086	22-23
Hepplewoodside NY9798	62	Hott NY7785	45
Herdlaw NY9498	61	Housey Crags NT9521	86
Hermitage NY9374	28	Houxty NY8578	42
Hesleyside Mill NY8084	44	Hudspeth NY9494	60
Hesleyside NY8183 NE48 2LA	43	Huel Crag NY8398	15
Heslop Crag NY7991	56	Huel Kirk NY8399	15
Het Hill NT9614	94	Hugh's Hill NY6971	38
Hetherington NY8278	41	Humble Burn	47
Hethpool Bell NT9028	89	Humble Hill NY6481	45
Hethpool Lake NT8928	90	Humble Law NY9697	61
Heugh Clints NY8780	27	Humbleton Hill NT9628	88
High Barns NY9267	34	Hummel Knowe NY7071	38
High Baulk NZ0070	31	Humshaugh NY9271	35
High Cantle NT9216	93	Hungry Law NT7406	10
High Carrick NY9296	59	Hunt Law NT9816	96
High Carriteth NY7983	44	Hunter's Burn	54
High Carry House NY8679	26	Hunter's Loap NT7206	10
High Farnham NT9602	73	Ill Sike	49
High Hawkhope NY7188	53	Ingram Glidders NT9916	96
High Long House NY6086	49	Ingram NU0116	96
High Moralee NY8476	37	Inner Hill NT8707	68
High Nick NY9387	23	Inner Hill NT8708	68, 81
High Pithouse NY9180	25	Inner Hill NT9111	80
High Rochester NY8398	15	Inner Hill NT9210	80
High Shaw NY9185	24	Inner Quickening Cleugh NT9016	93
High Shaw NY9498	61	Inner Strand	69
High Well House NY9774	29	Irthing	38
Highfield Burn	55	Jerry's Knowe	45
Highgreen Manor NY8091	56	Jerry's Linn NY7481	45
Highspoon Hill NT9001	64	Jock's Cleugh NY9496	60
Hillock NY8399	16	Jock's Crag NT7502	13
Hind Rigg NY8180	42	Johnside NY8088	56
Hindhaugh NY8784	24	Johnside Sike NY8188	56
Hindhope Law NY7697	16	Kateshaw Crag NT8707	68
Hindleysteel NY7472	38	Kateshaw Hill NY6586	48
Hindside Knowe NT8411	69	Keenshaw Burn	61
Hobb's Flow NY5690	49	Keeper's Cottage NY8191	56
Hog Knowe NT8308	68	Keepershield NY9072	36
Hog Lairs NT9113	80	Kelly's Pike NY8195	16
Hogdon Law NT9412	81	Kellyburn Hill NY8395	16
Hogswood Moor NY6795	52	Kelpie Strand NT9520	86
Hollows Hill NY7288	53	Kenterdale Hill NY9727	84
Holly Cleugh NT9524	85	Kerseycleugh NY6195	50
Holly Hall NY9766	32	Kettles, The Kettles NT9827	83
Holystone Burn	64	Kidland	79

Kidlandlee NT9109	79	Linnheads NY9386	23
Kielder Head NY6697	51	Linshiels Lake	67
Kielder NY6293	50	Linshiels NT8906	67
Kielder Stone NT6300	10, 51	Lishaw Rigg NY6086	48
Kielderstone Cleugh NT6400	10	Lisles Burn	22, 23
King's Chair	83	Little Bavington NY9880	29
Kingsley Crag NY7487	54	Little Buckster	39
Kingstone Dean NT9725	85	Little Burn	49
Kip Hill NZ0267	14	Little Cranecleugh Burn	47
Kip Law NY7150	14	Little Dod NT9514	94
Kirk Burn	41	Little Dodd NY7991	56
Kirk Hill NT9700	73	Little Hetha NT8828	90
Kirkfield NY8578	41	Little Swinburne NY9477 NE46 4TT	28
Kirknewton NT9130	89	Little Tosson NU0101	74
Kittyhirst NY6096	50	Little Wanny Crag NY9283	25
Knock Hill NT9916	96	Little Ward Law NT8614	69
Knocklaw NU0601	81	Little Whittington NY9969	31
Knocks, The Knocks NT8907	81	Loan Edge NT8005	14
Knockshield NY8350	81	Loaning Burn	60
Knox Knowe NT6502	10	Loft Cleugh	69
Kyloe Crags NY6983	46	Loft Hill NT8413	69
Lad's Clough NT7600	14	Loft Hill NT8725	90
Lady's Well NT9502	64	Long Crag NT8904	67
Ladyhill NY8075	38	Long Crag NT9617	94
Lake Wood	23	Long Crags NT9521	86
Lamb Crags NU1003	77	Long Crags NY6380	45
Lamb Hill NT8113	71	Long Crags NY6582	45
Lamb Hill NU1004	77	Long Hill NT8507	68
Lampert NY6874	38	Long Hill NY9099	64
Landshot NY9493	60	Long Rigg NY5885	48
Lanehead NY7985	57	Long Rigg NY6891	53
Langlee NT9623	85	Longheughshields NY8284	58
Langleeford NT9421	86	Longstrother NY8377	41
Lanternside Cleugh	63	Longtae Burn	64
Lanternside Edge NT9301	64	Lorbottle Burn	75
Latterford Doors NY8675	41	Lorbottle NU0306	75
Latterford NY8676	41	Lord's Shaw NY8291	21
Law, The Law NY6788	53	Lordenshaw NZ0598	77
Leadgate NY8177	40	Lough Crag NY9695	62
Leap Hill NT7207	10	Lough Knowe NY7395	54
Leaplish NY6587 NE48 1BT	48	Lough Shaw NY8489	21
Lee Hall NY8679	43	Loundon Hill NT9408	79
Leech Burn	85	Lounges Knowe NT8610	69
Leighton Hill NY9095	59	Lousey Law NY9278	29
Leonard's Hill NY8088	56	Low Alwinton NT9205	65
Lewis Burn	48	Low Barns NY9267	34
Lewisburn Colliery NY6388	49	Low Bleakhope NT9315	92
Lewisburn NY6590	48	Low Brunton NY9270	35
Liddell Hall NY9675	29	Low Cantle NT9315	92
Linbriggs NT8906	67	Low Carrick NY9195	59
Lincoln Hill NY9071	36	Low Carriteth NY7983	44
Linden Gill NZ1397	78	Low Carry House NY8578	26
Lindhope Burn	80	Low Cowden NY9178	27
Linen Sike	38	Low Cranecleugh NY6685	47
Linhope NT9616	95	Low Farham NT9702	73
Linhope Spout NT9517	95	Low Long House NY6287	49
Linn Moss	39	Low Moralee NY8476	37
Linn, The Linn NY7273	39	Low Park End NY8775	37

Low Roses Bower NY8075	40	Nether Houses NY8397	16
Low Shield NY8880	27	Nettlehope NT8911	80
Low Trewhitt NU0004	73	New Bewick NU0620	99
Lowstead NY8178	41	New Bingfield NY9873	30
Luckenarks NT9525	85	New Rift NY9466 NE46 4RW	33
Lumsdon Law NT7205	12	Newbiggin NY7889	55
Lyndhurst NY8675	37	Newtonrigg NY8375	37
Lynnholm NU0302	75	Newtown NU0300	77
Madam Law NT8626	90	Nichol's Pool NT8706	67
Maiden Castle	83	Nightfold Ridge NY8977	27
Main Stone NZ0298	74	Nishaw Burn	47
Makendon NT8009	71	North Barneystead NY8180	42
Mallow Burn	27	North Bridgford NY8482	43
Mally's Crag NT7900	14	North Middleton NT9924	84
Manor Farm NY8378	42	North Pike NT9613	94
Mantle Hill NY8184 NE48 2LB	43	North Riding NY9495	61
March Sike	72	North Yardhope NT9201	64
March Sike	94	Nunwick Mill NY8974 NE46 4BY	36
Marl Hill NY7776	39	Nunwick NY8775	36
Marven's Pike NY5787	49	Oh Me Edge NY7099	51
Matthew's Linn NY6490	48	Oh Me Sike NY6997	51
Meggrim's Knowe NT9615	94	Old Bewick NU0621	99
Memmer Kirk NT9212	80	Old Bridge End NY9265	32
Merry Burn	54	Old Hall NY7686	54
Mid Fell NY7489	53	Old Man's Sheel NY8083	44
Mid Hill NT8912	80	Old Quickening Cote NT8706	67
Mid Hill NT9021	91	Old Stell Crag NZ0398	74
Middle Cowden NY9178 NE48 3JB	27	Old Town NY8891	18
Middle Dean	96	Oldman Knowes NY7595	54
Middle Dean Camp NU0014	95	Otterburn NY8893	18
Middle Hill NT8712	69, 80	Ottercops Farm NY9588	19
Middle Moor NT9007	81	Ottercops Moss NY9589	19
Middle Woodburn House NY9086 NE48 2SG	23	Otterstone Lee NY6787	47
Middleton Hall NT9825	84	Outer Quickening Cleugh NT9017	93
Middleton Old Town NT9924	84	Ovenstone NY9698	61
Midge Hole NY9385	23	Overacres NY9093 NE19 1NA	59
Midge Holes NT9006	81	Padon / Peden Hill NY8192	10
Midgy Ha' NY9698	62	Padon Hill NY8192	20
Milkhope NT9211	80	Park Burn	60
Miller Burn NY8791	21	Park End NY8775	37
Millstone Crag NY6892	53	Park Head NY9085	24
Mindrum NT8432	17	Parkside NY8774	37
Monday Cleugh NT9528	88	Pass Peth	81
Monkridge Farm NY9191	18	Path Law NT8607	68
Monkridge Hall NY9092 NE19 1NB	18	Pathlaw Sike	68
Monkside NY6894	51	Pattenshiel Knowe NY9598	61
Mortley NY8277	41	Pauperhaugh NU0603	77
Mounces NY6588	48	Peat Sike	49
Mount Gilbert NY9489	19	Pedlar's Stone	61
Mount Pleasant NY7286	53	Peel Fell NY6299	10, 51
Mounthooly NT8822	91	Penchford NY9497	61
Mozie Law NT8315	70	Pete's Shank NT9710	71
Muckle Knowe NY6185	49	Philip's Cross NT7406	10
Mugsden Burn NU0717	99	Pinkie Shank NT9523	85
Nagshead Knowe NT9116	94	Pipers Cross NY6891	53
Neate Burn	49	Pit Burn	53
Needs Hill NY6690	52	Pit House NY8976	27
Nelly's Moss Lakes NU0802	77	Pit Houses NY8191	20, 56

Pithouse Crags NY6791	52	Ridge End NY7285	46
Planetrees NY9369	34	Riding, The Riding NY8284	58
Plashetts Carrs NY6791	53	Riding, The Riding NY9365	33
Plashetts NY6690	52	Ridlees Burn	67
Plashetts NY9681	29	Ridlees Hope NT8206	67
Pondicherry NU0401	76	Ridlees Road	67
Pope's Hill NY7389	53	Ridley Shiel NY7892	55
Port Gate	31	Ridley Stokoe NY7485	46
Portgate NY9868 NE46 4NF	31	Ridsdale NY9084	24
Potts Durtrees NY8797	17	Rimpside Hill NY9799	63
Pottstle Sike	81	Rimside Moor NU0806	63
Powburn NU0616	98	Rinds Burn	57
Prestwick Burn	26	Ringses, The Ringses NU0918	98
Prickly Knowe NT9813	95	Risey Burn	22
Pringle's Wood NU0718	99	Risingham NY8986	22
Pudding Burn	67	Ritto Hill NY9516	94
Puncherton NT9309	79	Riverhill Farm NY9073	36
Pundershaw NY7880	42	Rob's Pikes NY6899	51
Queen's Knowe NY6086	49	Robin Hood's Well NY9574	29
Queen's Sike	49	Robinson's Gears NY5783	48
Quickening Cote NT8806	67	Rochester NY8298 NE19 1RH	15
Ram's Haugh NT9205	65	Rockey's Hall NT9404	65
Ramsey's Burn	67	Rooken Edge NY7895	16
Ramshope Burn	12	Rooken Knowe NY8096	16
Ramshope Farm NT7304	12	Rookland Hill NT9308	79
Ramshope Lodge NT7204 NE19 1TZ	12	Rookling Law NT8506	67
Rattenraw Farm NY8595 NE19 1LH	20	Rose's Bower NY9971	30
Rattenraw NY8495	17	Roses Bower NY7975	40
Ravens Heugh NZ0198	74	Rothbury NU0501	76
Ravens Knowe NT7706	15	Rouchester Farm NY8977 NE48 3HT	27
Ravenscleugh NY9391	19	Roughside NY7483	45
Ravensheugh Crags NY8375	37	Roughting Linn NT9836	69
Raw Hill NT7700	13	Round Hill NY7477	40
Raw, The Raw NT7601	14	Round Top NY7176	40
Raw, The Raw NY9498	60-61	Routin Well NT8514	69
Ray Cleugh	95	Rowhope Burn	69
Ray Fell NY9585	22, 23	Rowhope NT8512	69
Ray Tongue NY9386	24	Rubbingstob Hill NY8978	27
Raylees Common	19	Ruken Sike	54
Raylees NY9291	19	Runners Burn	67
Reaveley Hill NU0018	95	Rushey Law NY9078	27
Reaver Crag NY9375	28	Rushey Rigg NY7075	39
Reavercrag NY9374	28	Rushy Gairs NT9424	86
Red Chirnells NU0302	76	Rushy Knowe NY6198	10
Rede	11	Rushy Knowe NY6588	48
Redesdale	11	Rushy Knowe NY6781	45
Redesdale Camp NY8299	15	Rushy Sike	94
Redesmouth NY8682 NE48 2ET	11	Russell's Cairn	70
Redeswire	11	Ryeclose Burn	54
Redewetter	11	Ryeclose NY7486	54
Redheugh NY7888	55	Ryehill NU0201	75
Redmire NY7985	57	Sadbury Hill NY8276	37
Redscar Sike	94	Saddler's Knowe NT8109	71
Reed Sike NY9380	25	Saddler's Slack NT9109	71
Reeker Pike NY6682	47	Salmonswell NY9466	33
Reenes Farm NY8284 NE48 2DU	21	Salter's Road	80-81
Rennies Burn	71	Sandhoe NY9766	32
Ridge End Burn	51	Sarelaw Crag NY9185	24

Saughieside NT8624	91	Smiddy Well Rigg NY8089	56
Saughy Hill NT8910	81	Snabdaugh NY7884	44
Scad Law NY7496	54	Snail Knowe NU0215	96
Scald Hill NY9221	86	Snear Hill NT9624	85
Scald Law NT8307	67	Sneep, The Sneep, The Sneap NY7988	56
Scaup Burn NT6500	51	Sneer Hill NT9011	80
Scaup NY6697	51	Snitter NU0203	75
Scaup Pikes NY6598	51	Snitter Windyside NU0104	75
Schil, The Schil NT8622	92	Snout End NT9516	94
Scotchcoultard NY7170	38	Snow Hall NY7986	57
Scotsman's Knowe NT9018	86, 93	Snuffies Scar NT9415	93
Scroggs, The Scroggs NT9700	73	Soldier's Fauld	63
Scroggs, The Scroggs NY8972	36	Sonsy Nook NY9385	24
Shank End NY6876	39	Sonsy Rigg NY6791	53
Shank, The Shank NT9622	85	Soot Burn	42
Sharperton NT9503	73	Soppit Farm NY9293	59
Shawdon Hill NU0813	98	South Middleton NU0023	84
Shaws, The Shaws NY8383	58	South Riding NY9495	60
Shawwell House NY9866	31	South Yardhope NT9200	64
Shedding Sike NT8918	93	Southernknowe NT8824	91
Sheel Law (Shieldlaw) NY8384 NE48 2HZ	21	Southhope Burn	67
Sheila Crag NY7976	40	Spithope Burn	12
Shellbraes NZ0071	31	Spout Hill NU0404	75
Shiel Crags Y8086	57	Spy Rigg NY6875	38
Shielcleugh Edge NT9217	92	St Gregory's Hill NT9129	88
Shieldfield NY8380	42	Stagshaw	31
Shilburn Haugh NY6986	46	Stagshaw Bank	31
Shill Moor NT9415	94	Stagshaw High House NY9767	31
Shilla Hill NY7690	55	Stagshaw House	31
Shillhope Law NT8709	68	Standard Hill NY8275	38
Shillmoor NT8807	68, 69	Standingstone Clints NY7976	40
Shipley Shiels NY7789	55	Standingstone Rigg NY8173	38
Shirlaw Pike NU1003	77	Stanegate	31, 34
Shitlington Hall NY8280	42	Staniel Heugh NY9187	23
Shittleheugh NY8694	17	Stannersburn NY7286	46
Shivers Cleugh NU0016	96	Steel, The Steel NY8982	24
Sidwood Cottage NY7789	55	Steely Crag NT9524	85
Sills Burn NT8202	15	Stewart Shiels NY8698	17
Silver Hill NY9467	34	Sticky Bog NT9423	86
Silverton Hill NT9308	79	Stiddlehill NY9185 NE48 2DD	23
Simonburn Castle NY8673	36	Sting Head NY9312	81
Simonburn NY8873	36	Stobbs Farm NY8397	16
Simonside NZ0298	74	Stock Brook	99
Sinkside Hill NT8826	90	Stokoe High Crags NY7584	46
Skirl Naked NT9725	85	Stokoe NY7386	46
Slade Sike	42	Stone House NY8180	42
Slaterfield NY8674	37	Stonehaugh NY7976	40
Slime Foot	69	Stony Sike	49
Slippery Crags NT8804	67	Stooprigg NY8472	37
Smalehope Sike	51	Story's Gairs NY6282	48
Smales Burn	46	Stott Crags NY5983	48
Smales Leap	46	Stower Hill NY6485	47
Smales NY7184	46	Street Head NY7398	54
Smalesmouth NY7385	46	Street, The Street	69
Small Burn	38, 67	Struther Bog NY8195	16
Smallhope Burn	55	Stuckin Knowe NY5991	49
Smallhope Sike	51	Summit Cottages NY9384	22
Smeddum Sike NT8721	92	Sunday Burn	38

Sundaysight NY8189	56	Troughend NY8591 NE19 1LA	20
Sutherland Bridge NT8824	91	Trouting, The NT6502	10
Swallow Knowe NU0705	77	Trows Law NT8513	69
Sweethope NY9581	25	Trows Road End	69
Swin Burn	28	Trows, The Trows NT8512	69
Swinburne Castle NY9375	28	Trows, The Trows NT9527	84
Swindon NY9799	62	Trowupburn NT8976	90
Swine Hill NT9082	24	Turf Knowe NU0015	96
Swineshaw Burn	43	Turfy Knowe NY6097	10
Swineside Law NT8313	69	Tyne, River Tyne	33
Switcher Wood NT9725	85	Upper Stony Holes NT6501	51
Tarn Burn	54	Uppertown NY8672	36
Tarret Burn	6	Uswayford NT8814	80
Tarset Burn NY7397	54	Wadge Head NY7985	57
Tarset Castle NY7985	57	Wagtail Farm NU0700	77
Tate's Well NT7304	12	Wainfordrigg NY9196	59
Tathey Crags NT9621	86	Wainhope NY6792	52
Tecket NY8672	36	Walkerwalls NT9825	84
Thatchy Sike	46	Wall NY9168	35
The Eals NY7685	45	Walwick Grange NY9069	34
The Forks NY6388	49	Walwick NY9070	34
The Goatstones	37	Wanney Byre NY9383	25
The Knares NY6480	45	Wanney, Wannies	25
The Trinket NY7577	40	Wansbeck	25
Thirl Moor NT8008	71	Ward Law NT8613	69
Thistle Crag NY9288	23	Warden NY9166	34
Thistle Riggs Farm NY9167	34	Wark, Wark on Tyne NY8677	41
Thockrington NY9578	29	Warks Burn	39
Thorney Burn	54	Warksburn	41
Thorneyburn NY7686	54	Warksfield Head NY8478	42
Thorneyhirst NY8678	27	Warkshaugh Bank NY8677	27
Three Pikes NY6695	52	Warkswood NY8478	42
Threeburn Mouths NY9486	23	Warton NU0002	75
Thropton NU0302	75, 76	Watch Crags NY7882	43
Thross Burn	39	Watch Grain	49
Through Hill NT8607	68	Watch Hill NT9526	84
Thrum Mill NU0601	77	Watch Hill NT9825	84
Tick Law NU0821	99	Watch Trees NY6860	39
Tilesheds NY8592	20	Watergate Farm NY8179	42
Tipalt Burn	39	Waterside Plantation NU0719	99
Titlington NU1015	98	Watling Street	9
Tod Crag NY9387	23	Watson's Walls NY7881	42
Tod Knowe NY7699	13	Watty Bell's Cairn NT8901	64
Tod Law NT7700	13	Waud House NT9827	83
Tod Law NY8397	13, 16	Wedder Hill NT7911	71
Todridge NY9971	30	Weldon NZ1398	78
Tofts Burn NY8791	20	Well House NY9674	29
Tofts NY8592	20	Wellhaugh NY6689	48
Tone Inn NY9180	25	West Hepple NT9700	73
Tone Thrasher NY9079	25	West Highridge NY8180	42
Tongue, The Tongue NT8724	91	West Hill NT9029	88
Torleehouse NT9128	89	West Hill NU0214	97
Tosson Hill NZ0098	74	West Wilkwood NT8703	67
Townhead NY8774	37	West Wood NT9205	65
Townshield Bank NY8172	37	West Woodburn NY8986	22
Trewhitt Moor NT9804	74	Wester Dean	85
Trewhitt Steads NU0006	74	Wester Hall NY9172	35
Trewhitt, High Trewhitt NU0005	73	Wester Tor NT9027	88

Westfield House NY9267	34	Windyhaugh NT8610	69
Westnewton NT9030	89	Winter's Gibbet NY9690	60
Wether Cairn NT9411	80	Wirchet NY6297	51
Wether Hill Camp NU0015	96	Wishaw NY9487	23
Wether Hill NT9012	80	Witch Crags NT8705	67
Wether Hill NY9290	19	Witchy Nick NY9899	62
Wether Lair NY7097	51	Woden Law NT7612	9
Wether Lair NY7695	55	Wolf Hole NU0903	77
Wether Law NY7695	16	Wolfershiel NU0100	74
Wetshaw Hope NY8589	21	Wood Hall NT9503	64
Wetshaw Sike NY8689	21	Woodburnhill NY9186	22
Whar Moor NT8311	70	Woodhall Haugh	64
Whetstone House NY9286	23	Woodhall Wood	64
Whickhope Nick NY6681	45	Woodhead NY8183	43
Whickhope NY6886	46	Woodhill NY8892	18
Whiggs, The Whiggs NY9970	31	Woodhouse NY8887 NE48 2SY	22
Whinny Hill NY6889	53	Woodley Shield NY8476	41
Whitchester NY7783	44	Woodpark NY8479	42
White Crag NY6790	53	Woody Crags NY6898	51
White Crags NT6901	12	Wool Meath NY7099	13
White Fell NY6190	49	Woolaw Farm NY8298	15
White Hill NY7488	54	Woolaw NY8298	13
White Hill NY7776	39	Woolbist Law NT8207	67
White House NY9280	25	Wooler NT9928	83
White Kielder Burn	51	Wooler Water	84
White Meadows NY9487	23	Woolmeath Edge NY7199	13
White Riggs NZ0073	31	Wreigh Burn	73
White Side NY7185	46	Wreighill NT9701	73
White Side NY7576	39	Wreighill Pike NT9802	73
Whitefield Camp	62	Written Crag NY9368	34
Whitefield Edge NU0803	77	Wylies Craigs NT6401	10
Whitehall NT8826	90	Yarnspath Law NT8813	80
Whiteheugh Crag NY7494	55	Yarrow Moor NY7086	46
Whitehill Moor NY6978	39	Yarrow NY7187	46
Whitehill NY6777	39	Yarrowmoor NY7086	46
Whitelee Knowe NT8803	67	Yatesfield Farm NY8697	17
Whitelee NT7105 NE19 1TJ	12	Yatesfield Hill NY8597	17
Whitelee Sike	66	Yearning Hall NT8212	71
Whiteside NY9180	25	Yearning Law NT8211	70
Whitewell Strand	92	Yeavering Bell NT9229	58
Whitley Pike NY8291	20	Yeavering Bell NT9229	89
Whittington White House NZ0172	31	Yett Burn	49
Whittle NU0204	75	Yoke Burn	80
Whitton NU0501	77		
Wholhope Hill NT9311	80		
Wholme Gill NZ1194	78		
Whygate NY7776	39		
Wideopen	91		
Wideopen Head NT8626	90		
Wiley Sike NY6740	10		
Wilkwood Burn	67		
Willowbog NY5989	49		
Willowbog NY7975	38		
Willy Crag NY9288	23		
Wind Burn	16		
Wind Hill NY6888	53		
Windy Edge	40		
Windy Gyle NT8515	69, 70		